Our Geologic Heritage in Colorado's Northern Front Range Foothills

A Guide to Larimer County Natural Areas

Michael B Kendrick

ATTENTION CORPORATIONS, UNIVERSITIES, COLLEGES, AND PROFESSIONAL ORGANIZATIONS:
Quantity discounts are available on bulk purchases of this book for educational or gift purposes. Special books or book excerpts can also be created to fit specific needs. Please contact kendrickm970@gmail.com for inquiries.

Perhaps no finer locality exists in the West for the careful study of the different sedimentary formations and their relations to the metamorphic rocks than along the overland stage road from Laramie to Denver—F.V. Hayden

Previously Unpublished Sketches by Henry W. Elliot united with the Preliminary Field Report of the United States Geological Survey of Colorado and New Mexico, 1869, by F.V. Hayden

PREFACE

It is a fortunate circumstance that within Colorado's Front Range foothills, geologic processes have exposed such a rich historical record about how our local region developed into how we know it today. And just as importantly, the foothill communities have taken it upon themselves to become stewards of this landscape by protecting much of it as publicly accessible natural areas.*

Our sense of place and home within this landscape is deepened when we go beyond scenic visual aesthetics to discover and understand the geology of the area. After all, it is geologic processes that determine such things as topography, elevation, climate, drainage, and mineral resources. In turn these determine the environments which foster flora, fauna, and human communities.

Many people are curious about rock types, their age, what they tell us, how to identify them, what uses they have, fossils they contain, geologic hazards they pose, and how they are mapped within our natural areas.

This guide is meant to inform and educate with a focus on the geologic features and rocks exposed in the natural areas of the Northern Colorado Front Range foothills of Larimer County. The information herein is sourced from industry, academia, government, and popular publications as well as time spent observing on the trails.

The guide is aimed at all who wish to learn about our local geologic neighborhood. It is a brief summary designed to provide comprehensive coverage, yet thin enough to be tucked into a backpack and taken along on your foothills journeys.

Special thanks to Frank Ethridge (Professor Emeritus of Geology, Colorado State University) for his helpful comments and suggestions. My appreciation also to Ethan Foose of Laurel Editing, as well as readers of earlier drafts including Joyce Turley, Pat Hayward and Macy Reynolds.

Finally, a big thank you to Rick Turley for assistance with photography and my wife Shirley for her pen&ink/watercolor renderings of the schematic rock column figures.

*The term "natural areas" will be used in a broad referential sense to include the City of Fort Collins Natural Areas, Larimer County Open Spaces, Lory State Park, and Watson Lake State Wildlife Area.

CONTENTS

PART D: GEOLOGIC GUIDE TO SELECT NATURAL AREA HIKES

INTRODUCTION

Extensive trails within Larimer County's foothills natural areas provide year round access to scenic landscapes harboring a wide diversity of plant and animal life. The landscape is a reflection of the underlying geology revealed by widespread rock exposures. In a short span of a few miles, a succession of tilted sedimentary rocks rising from the Great Plains lead to lower montane elevations where metamorphic and igneous rocks mark the beginning of Colorado geologic history.

Whether just beginning on your geologic journey of discovery, or are a seasoned geology enthusiast, this guide is a resource towards learning or acquiring a deeper geologic understanding and appreciation of the foothills. The guide is comprehensive in coverage, yet serves as a jumping off point towards further exploration of the many topics covered.

While simply reading the guide will be informative, it will be quite satisfying to get out on the trail and see the geology in person. I find, and hope you will too, a deeper connection with the landscape that comes with knowing the geologic history of the ground we walk and live upon.

This introduction is a short primer with useful reference figures as background for this guide. In particular, the geologic **cross-section** and **stratigraphic chart** are fundamental references for knowing the **rock unit** names, ages, topographic expression, and locations with respect to the natural areas. Throughout the guide, geologic terms in bold font refer to definitions provided in the glossary. The document is full of figures and photos. Many of the figures are hand-drawn, others are modified from the literature.

In Part A, a summary of each of the rock units exposed in the Northern Colorado **Front Range** foothills is broken up into three sections: the first captures basic facts on age, thickness, **lithology,** and **depositional environment**; the second identifies some of the best natural areas and trails for observing a specific rock unit, including tips on how to identify them. The third section, Topics of Interest, provides background on topics such as resource use, relevant geologic events, explanations of conspicuous rock features and fossils, as well as geologic hazards. The summaries sections start with the oldest rocks, the Precambrian Crust, and sequentially works up to the youngest formation, the Pierre Shale.

Two short sections follow: Part B is a discussion of the Laramide Orogeny, the mountain-building event that created the Rocky Mountains and affected all the rocks reviewed in Part A, and Part C focuses on The Soapstone Prairie Natural Area, where the geologic story picks up after the Laramide Orogeny.

Finally in Part D, twelve hikes have been chosen to highlight the full range of rock types, formations, and geologic features exposed in our Northern Colorado foothills natural areas. For each hike, a short summary, trail map with geologic overlay, and key pictures are provided. For more geologic detail, the hiker should refer back to relevant sections in Part A.

Before delving in to Part A, it is helpful to clarify naming conventions. The term 'rock unit' is generic and field geologists have come to use a set of clearer definitions. In the developed hierarchy, a **formation** (fm) is the fundamental unit and refers to a recognizable interval of rock that can be mapped over a fairly broad region and is confined to a distinct interval of time.

Some formations are made up of more than one rock type, for example, shale and sandstone, while others may be of a single rock type. In the latter case the rock type may be incorporated in the name—for example, Lyons Sandstone.

All the **sedimentary rock** units in the foothills have formation status except the Dakota and Benton, which are recognized as a collection of formations and therefore have a higher **group** status. And, on a finer scale subdivision, some formations contain **member** and 'sub-member' intervals. Where this is the case it is annotated on the schematic rock column figures in each summary.

When referring to rocks in general, regardless of the hierarchical status, the term 'rock unit' will still be used, otherwise when discussing a particular rock unit it will be referred to with respect to its formation, group, or member status.

Lastly, throughout the text there are abbreviated geologic ages of rocks and events dated millions or a billion plus years ago that are designated *mya* and *bya* respectively. Sometimes, when speaking of a geologic time interval, these may be shortened to *my* or *by.* For example, the Great Unconformity represents a gap in the rock record of greater than 1 by.

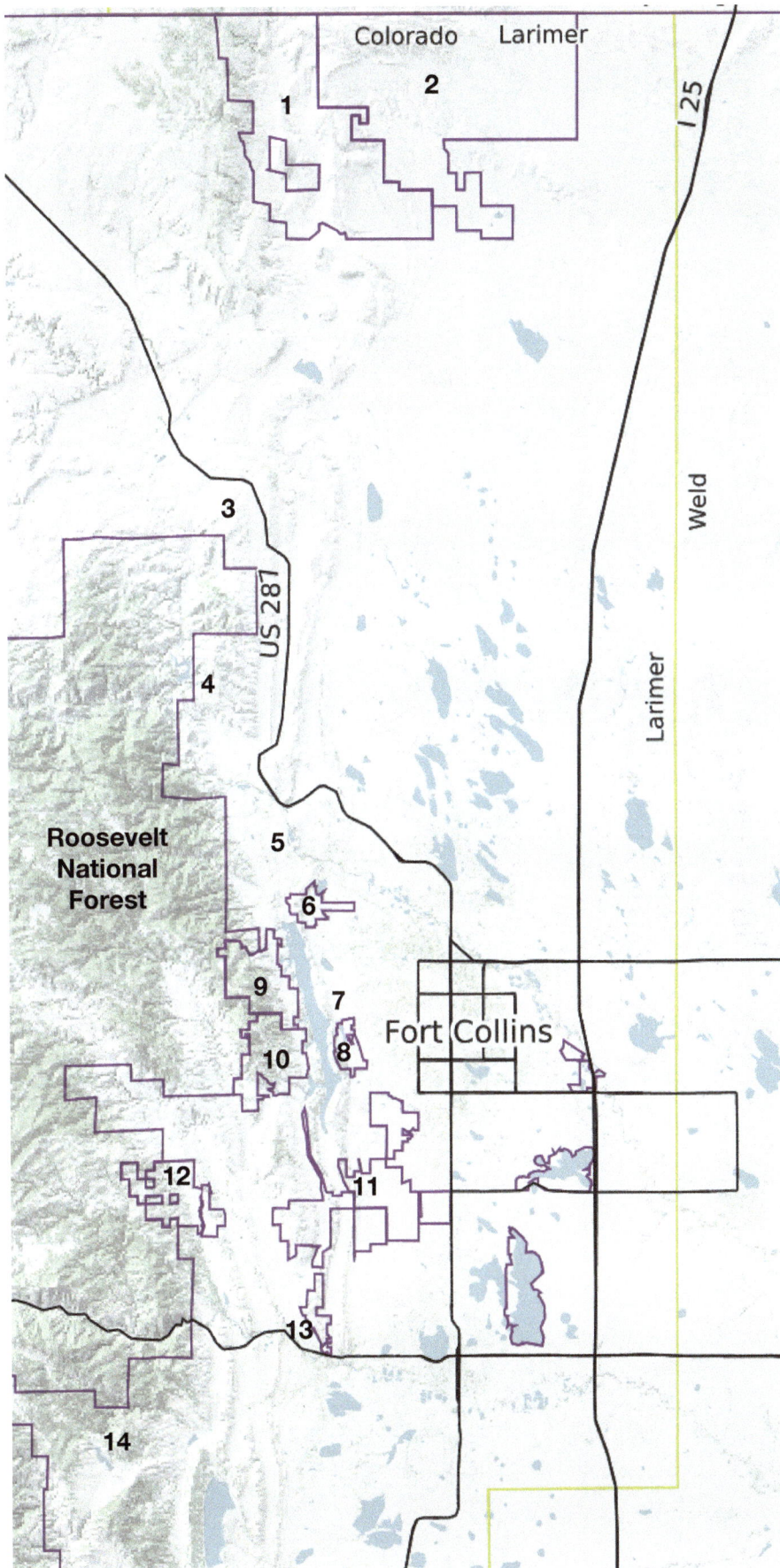

Fig. 1 Natural Areas Locations

1 Red Mountain Open Space
2 Soapstone Prairie Natural Area
3 Eagle's Nest Open Space
4 Gateway Natural Area
5 Watson Lake State Wildlife Area
6 Reservoir Ridge Natural Area
7 Maxwell Natural Area
8 Pineridge Natural Area
9 Lory State Park
10 Horsetooth Mtn. Open Space
11 Coyote Ridge Natural Area
12 Bobcat Ridge Natural Area
13 Devil's Backbone Open Space
14 Ramsay-Shockey Open Space

Note: The term "natural areas" is used throughout this guide to include Fort Collins Natural Areas, Larimer County Open Spaces, Lory State Park, and Watson Lake State Wildlife Area. Not all natural area outlines shown on map.

Basemap copyright by OpenStreetMap contributors

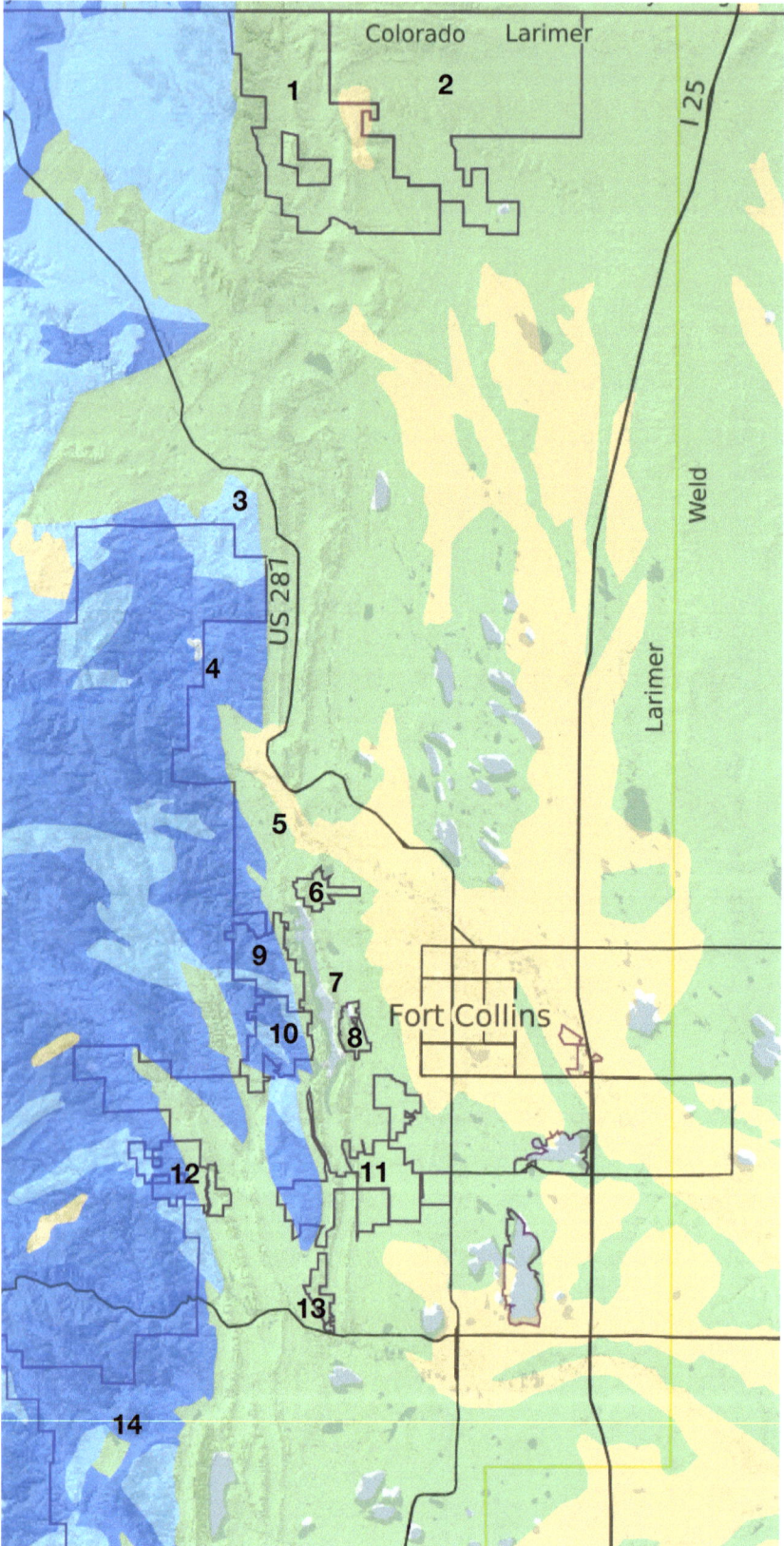

Fig. 2 Simplified Geologic Map
with Natural Area Locations

Green - Sedimentary Rock

Light Blue - Igneous Rock

Dark Blue - Metamorphic Rock

Yellow - Unconsolidated Sediments

See Fig.1 for natural area
names of numbered locations

Basemap copyright by OpenStreetMap contributors

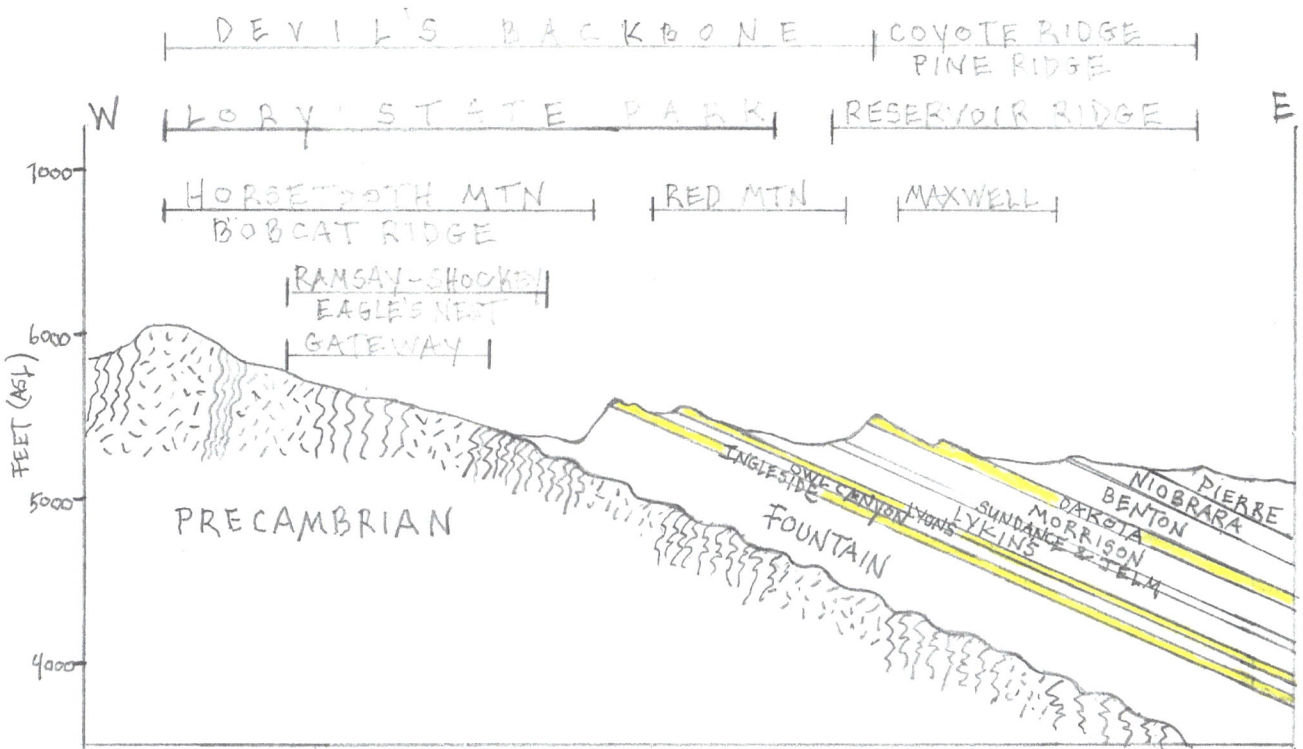

Fig. 3 Geologic Cross Section. Typical topographic expression and subsurface structure of the rock units in Northern Colorado's Front Range foothills. The colored rock units highlight the conspicuous ridges known as **hogbacks**. From east to west they are: the Dakota, Lyons, and Ingleside. Horsetooth Reservoir occupies the valley underlain by the Lykins Formation. The bars labeled with Natural Areas over the cross section represent their geologic boundaries with respect to trail access. The rock units are progressively older from east to west.

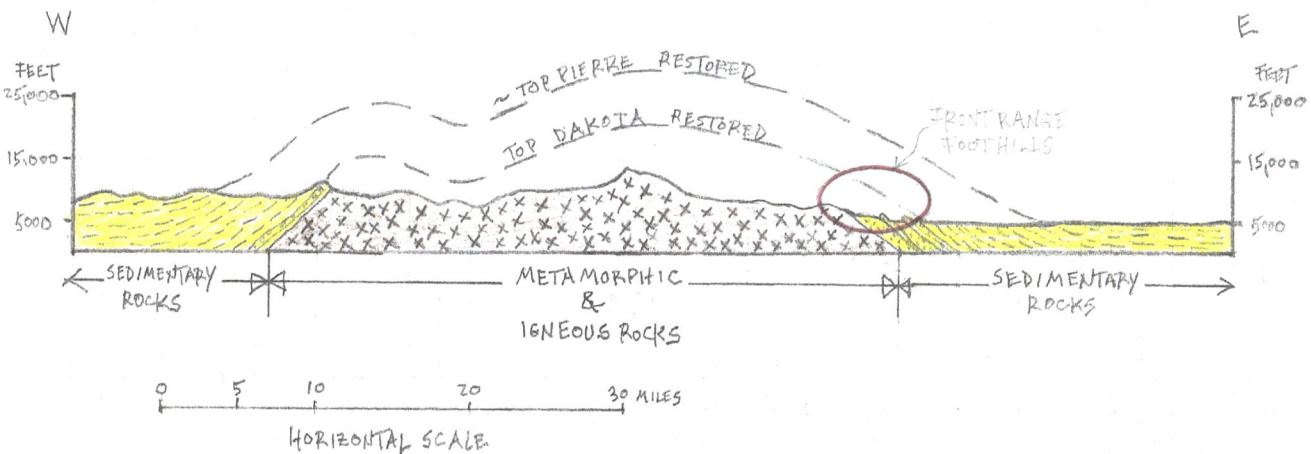

Fig. 4 Regional Cross Section. Cross section across the Front Range with the location of the foothills discussed in this guide. (Modified from Chronic, 1972)

ERA | **PERIOD** | **AGE (mya)** | **ROCK UNITS** | **MAJOR GAPS IN ROCK RECORD** | **GEOLOGIC EVENTS** | **GEOLOGIC SPAN OF NATURAL AREAS**

MESOZOIC — CRETACEOUS:
- 70 — FOX HILLS SANDSTONE
- 80 — PIERRE
- 90 — NIOBRARA
- BENTON
- 100 — DAKOTA
- 102

MESOZOIC — JURASSIC:
- 148 — MORRISON — ~46 MY
- 157
- 165 — SUNDANCE — ~8 MY
- 170

MESOZOIC — TRIASSIC:
- 240 — JELM — ~70 MY
- 250

PALEOZOIC — PERMIAN:
- 260 — LYKINS
- 270
- 280 — LYONS / OWL CANYON
- INGLESIDE
- 290

PALEOZOIC — PENNSYLVANIAN:
- 300 — FOUNTAIN
- 310
- 1400 — GREAT UNCONFORMITY — >1 BY

PRECAMBRIAN:
- ~1.4–1.7 BYA IGNEOUS INTRUSIONS
- ~1.8 BYA SCHISTS & GNEISS

GEOLOGIC EVENTS:
- WESTERN INTERIOR SEAWAY
- SEVIER OROGENY
- AGE OF DINOSAURS
- PANGAEA SUPERCONTINENT
- PERMIAN EXTINCTION
- ANCESTRAL ROCKIES
- FORMATION OF COLORADO CRUST
- ACCRETION TO WYOMING CRATON

GEOLOGIC SPAN OF NATURAL AREAS:
- COYOTE RIDGE
- PINE RIDGE
- RESERVOIR RIDGE
- MAXWELL
- DEVIL'S BACKBONE
- LORY STATE PARK
- RED MOUNTAIN
- HORSETOOTH MTN
- BOBCAT RIDGE
- RAMSAY–SHOCKEY
- EAGLES NEST
- GATEWAY

Fig. 5 Stratigraphic Chart. This chart summarizes the vertical succession of rock units in the Northern Colorado Front Range foothills. The age of the rock units are labeled on the left. The geologic events in the middle column are discussed in the text. All these rock units were created before the rise of the modern Rocky Mountains.

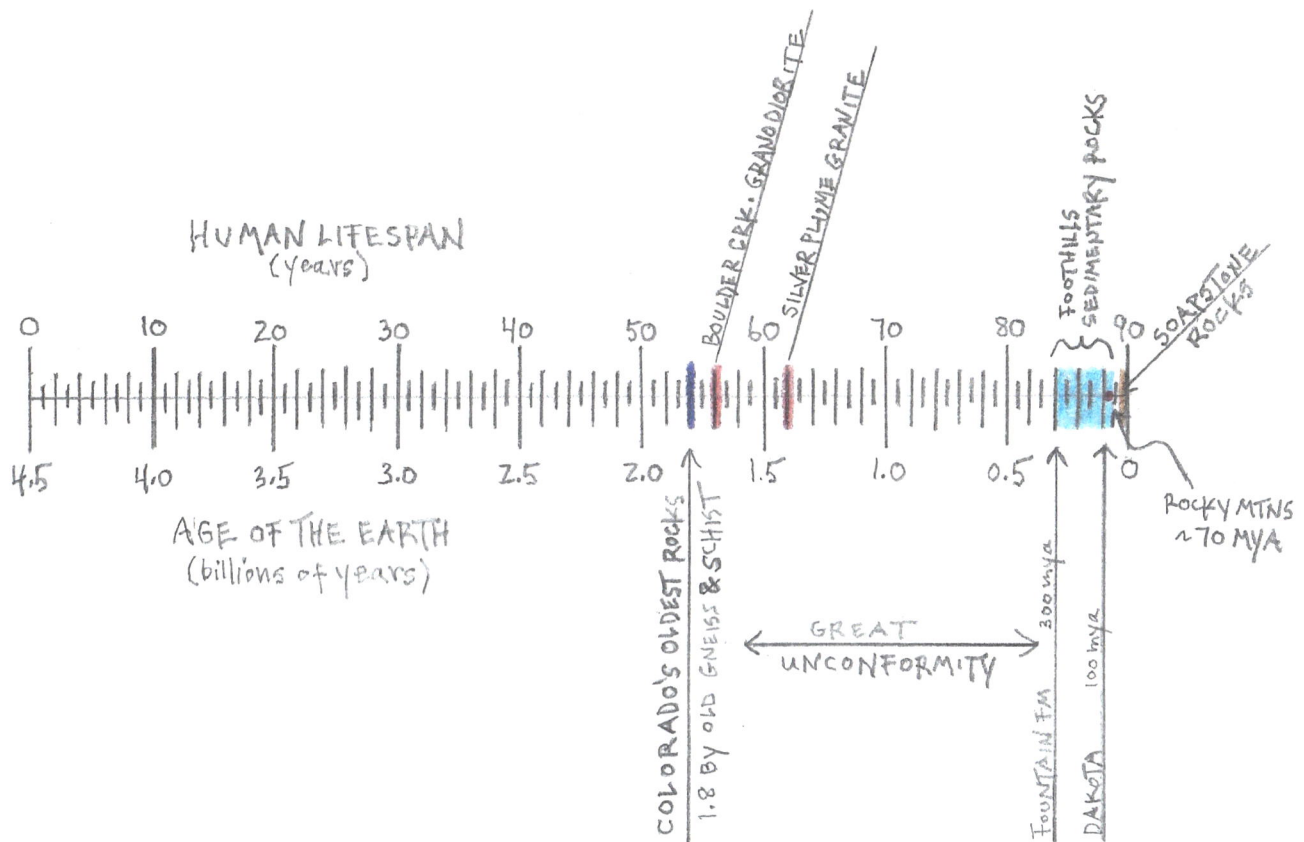

Fig. 6 Fathoming Geologic Time. Color bands are ages of rocks seen in the Front Range foothills. Metamorphic schist and gneiss are the oldest rocks dated at circa 1.8 billion years old. They were intruded by two igneous events dated at circa 1.7 and 1.4 billion years, respectively. There then follows a large gap in time (> 1 by) known as the Great Unconformity for which there is no rock record. The blue colored band are the tilted sedimentary rocks that rise from the Great Plains and are dated from circa 300 million years old (the Fountain Formation) to 70 million years old (the Pierre Shale). After the Rocky Mountains were uplifted in the period from 70—40 million years ago, the relatively flat lying rocks exposed in Soapstone Prairie NA were deposited from approximately 36—5 million years ago.

To add some perspective to geologic time, a human lifespan of 90 years is scaled so that one human year equals 50 million earth years. Using the rock record as event markers in this imagined 90 year lifespan, there is no memory of anything that happened before age 54. And again, between the ages of 62 and 84 (the Great Unconformity) no memory of events is recalled. The Rocky Mountains would have begun to form after the age of 88. The obvious point here is the fragmentary record and relative recentness of the geologic record with respect to the great age of the earth.

7

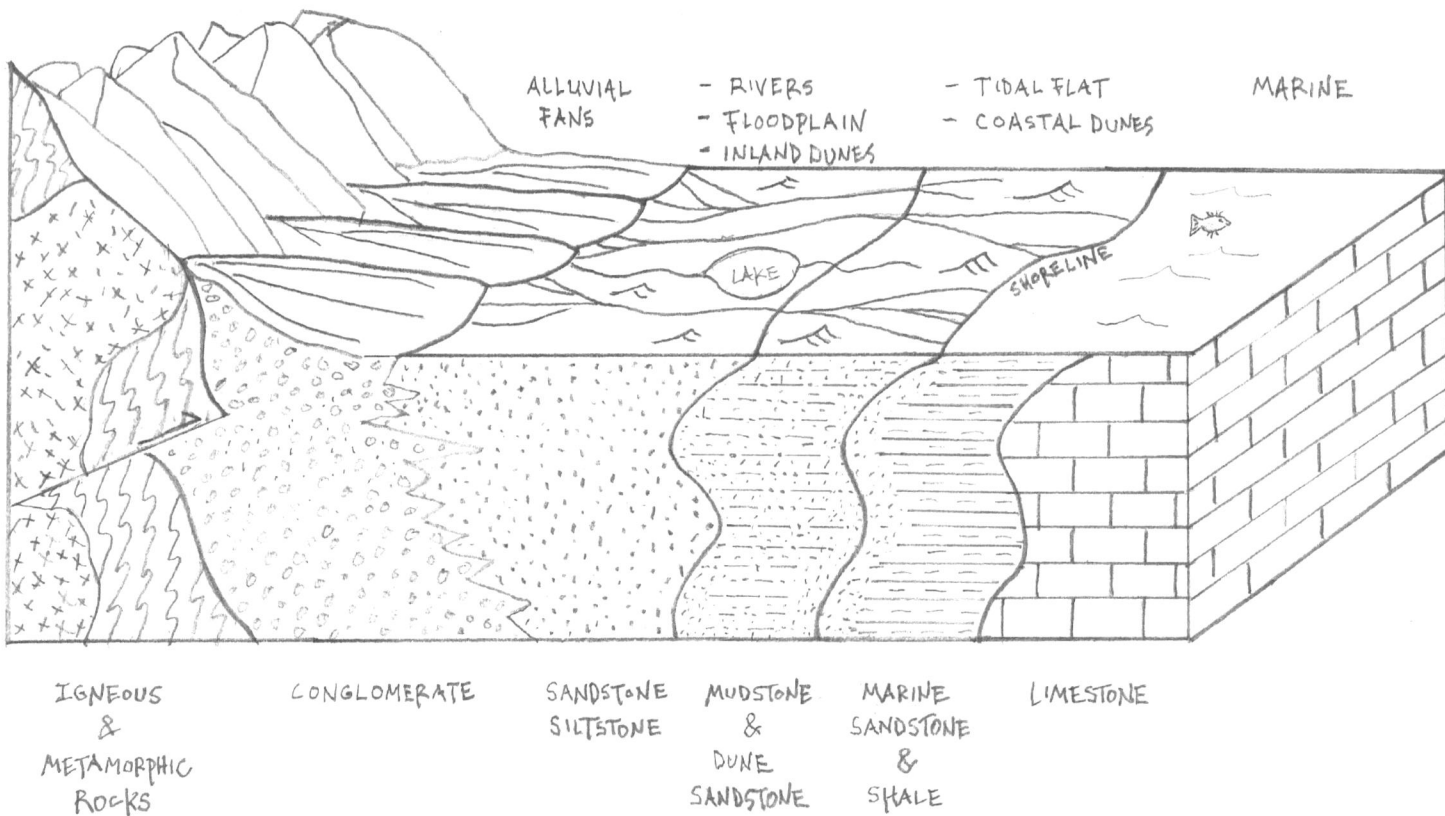

Fig. 7 Environments of Deposition. The sedimentary rocks of the northern Colorado Front Range foothills were deposited in a range of environments represented in this block diagram. For example, a significant part of the Fountain Formation was deposited as alluvial fan and braided stream deposits adjacent to the Ancestral Rocky Mountains, while the Pierre Shale was deposited within the marine Western Interior Seaway. Some rock units were deposited in one environment, for example, the coastal dunes of the Lyons Sandstone, while others spanned a range of environments such as the Dakota Group which includes both river and marine deposits. The summaries in Part A of this guide refer to each rock units depositional environment and corresponding rock type. (Modified from Huntoon et al., 2014)

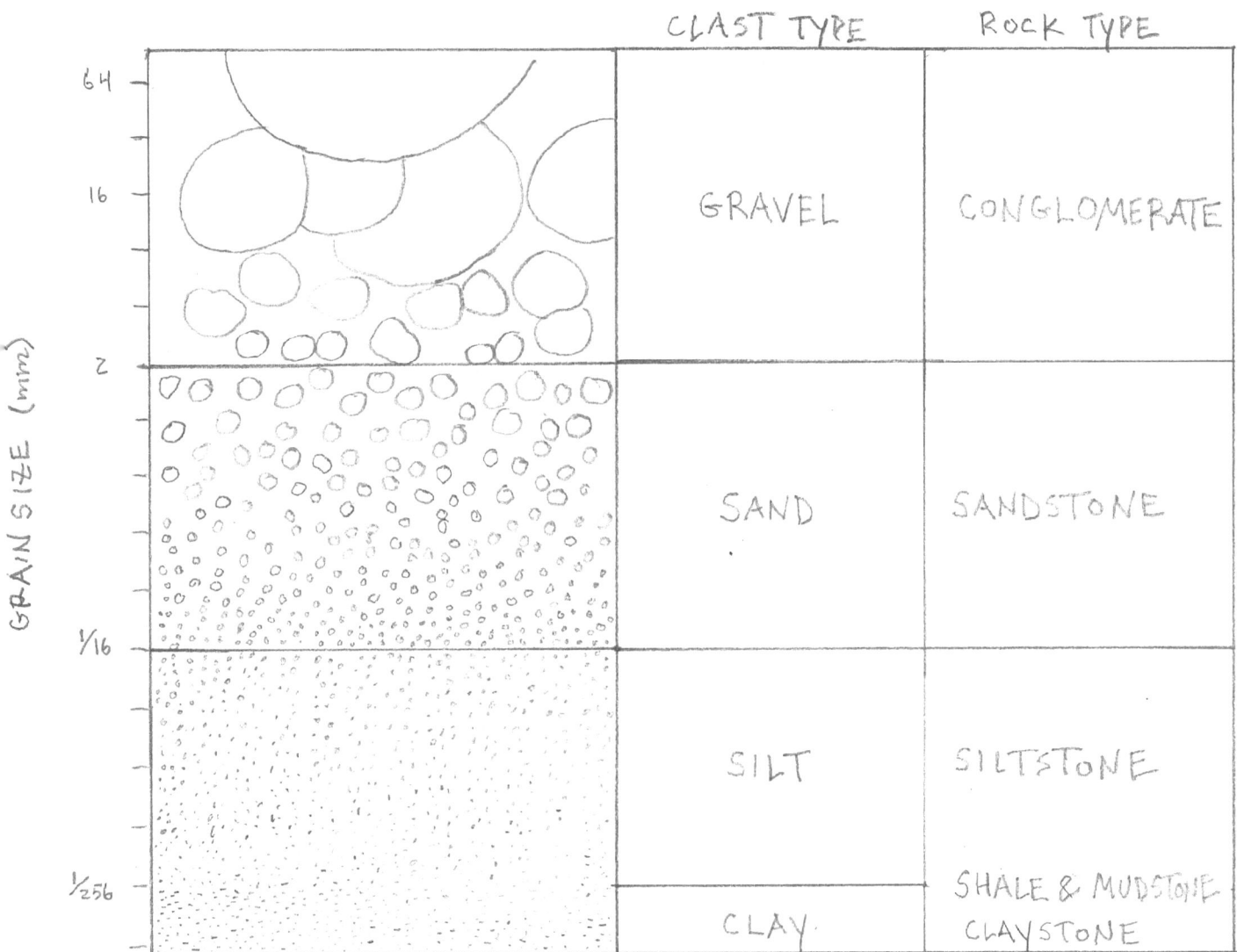

	CLAST TYPE	ROCK TYPE
GRAVEL	CONGLOMERATE
SAND	SANDSTONE
SILT	SILTSTONE
CLAY	SHALE & MUDSTONE CLAYSTONE

GRAIN SIZE (mm)

64
16
2
1/16
1/256

Fig. 8 Classification of Clastic Sedimentary Rocks. The vast majority of sedimentary rocks in the foothills are **clastics**. They are composed of broken fragments of pre-existing rocks from which they were weathered and eroded. The names assigned to the loose sediments and the rocks they form is based only on grain size, not composition. Sand deposits become sandstone, gravel deposits become conglomerate etc. (Modified from Earth's Dynamic Systems,10th Ed. Prentice-Hall)

Other rock types not as volumetrically significant as the clastic rocks in the foothills are **limestone** and **gypsum**. Limestones are typically formed by biogenic or chemical precipitation of calcium carbonate ($CaCO_3$) while gypsum ($CaSO_4$ $2H_2O$) precipitates as concentrated salts lose their solubility when lake or sea water evaporates. **Marlstone** is a combination of clay and silt sized clastics mixed with limestone.

9

	Building/ Landscape Stone	Aggregate	Industrial/ Manufactoring	Mineral Ore	Oil & Gas	Notable Fossils	Comments
Pierre					X	X	Sandstone members produce O&G in Denver Basin Ammonite fossils
Niobrara			X		X	X	Ft. Hayes Limestone mined for concrete industry Smoky Hill chalks produce O&G in Denver Basin Inoceramus clam fossils
Benton					X	X	Codell Ss & Greenhorn Ls produce O&G in Denver Basin Inoceramus clam fossils
Dakota	X		X		X	X	Active quarry near Bellvue for landscape rock Major O&G producer in Denver Basin (Muddy Ss) Historic claystone mines near Golden, CO Dinosaur trackway at Dinosaur Ridge near Golden, CO
Morrison				X		X	Uranium/Vanadium mines in SW CO. Dinosaur Bones near town of Morrison
Sundance							
Jelm							
Lykins			X			X	Active Gypsum quarry near Livermore, CO Stromatolite fossils in Forelle Limestone
Lyons	X				X		Active quarries near Masonville and Lyons, CO Historic quarries at Horsetooth Reservoir Minor O&G production in Denver Basin
Owl Canyon							
Ingleside	X	X	X				Active quarries near Owl Canyon Historic quarries at Horsetooth Reservoir
Fountain							
Precambrian	X			X			Granite 'moss rock' for landscaping Historic gold mines near Rustic, CO

Fig. 9 **Natural Resource Table**. Historic and current natural resources associated with the rocks of the Northern Colorado Front Range foothills. Oil and gas extraction in the Denver Basin, and surface quarrying operations in the foothills are ongoing activities. (Derived in part from Reno WR (2008) Front Range Sandstone Quarries. University of Colorado at Boulder)

PART A

Geologic Overview

of the

Rock Units

Aerial photo looking north toward Horsetooth Reservoir. The Rimrock Trail in Devil's Backbone Open Space is visible towards the bottom of the photograph. Each of the labelled rock units are addressed in this section of the guide.

PRECAMBRIAN CRUST

AGE: circa 1.8 – 1.4 billion years ago

THICKNESS: Approximately 30 miles (estimated from seismic data)

ENVIRONMENT of FORMATION: Mountain roots, convergent plate margin

LITHOLOGY: Metamorphic schist, gneiss, and igneous intrusions

---

WHERE ARE GOOD EXPOSURES OF THE PRECAMBRIAN?

Eagle's Nest:
  Three Bar and OT Trails - Exposure of Sherman Granite
Gateway:
  Black Powder Trail - Exposure of mica schist, gneiss, and migmatite
Lory State Park:
  Arthur's Rock Trail - Exposure of Boulder Creek Granodiorite, pegmatite, and schist
  Well Gulch Trail - Exposure of mica schist and gneiss
Horsetooth Mountain:
  Horsetooth Rock Trail - Schist in lower reaches but mostly pegmatite
Bobcat Ridge:
  Power Line Trail - Granitic (tonalite) outcrops at the higher elevations
  Valley Loop Trail - Schists on the west segment of trail
Devil's Backbone:
  Indian Springs Trail - Exposure of schist and gneiss
Ramsey-Shockey:
  Shoshone Trail - Exposure of mica schist

HOW DO I IDENTIFY THE PRECAMBRIAN?

- Look for **foliation** in metamorphic rocks: flat, parallel aligned mica layers (schist) or alternating light and dark colored banding (gneiss). In outcrops the schists often have a tombstone appearance

- Granitic rocks display visible interlocking mineral crystals (e.g., feldspar, quartz, mica). Commonly rounded or smooth weathered in outcrop

- Pegmatites are very coarse textured with mineral crystals exceeding 1 inch in size

---

TOPICS OF INTEREST:

Ancient Mountains

The **schist**, **gneiss** and **granitic** rocks exposed in the Northern Colorado Front Range foothills are the roots of an ancient mountain belt created by the collision of two landmasses. Just as **plate tectonic** motion drives India into Asia, creating the modern Himalayas, a collision about 1.8 billion years ago, known as the **Yavapai Orogeny**, created the mountain root core of rock that is the **basement** foundation of Colorado. In the Medicine Bow Range in Wyoming, geologists have mapped and named the contact line between the ancient colliding landmasses as the Cheyenne Belt. The rocks north of the Belt are more than two billion years old and part of the older Wyoming Craton, while south of the Belt, and extending across Colorado, the oldest rocks are approximately 1.8 billion years old.

Metamorphic Rocks in Colorado's Front Range

Thrust faults are part of the mountain building process where one block of rock moves over the top of another along a fault. As fault blocks progressively stack one on top of another a mountain range grows, but simultaneously rocks are also buried to great depths beneath the succession of thrust faults. When the burial depths reach about six miles, heat and pressure begin to change the mineral assemblages and textures of the original rocks. This process is known as metamorphism. Typically, with increasing depth there is an established progression. For example, an original marine shale will transform initially into a slate, then a phyllite, then a schist, a gneiss, and finally through partial melting into a **migmatite**.

It is estimated that the **metamorphic rocks** we see today in the Front Range foothills were once buried up to a depth of 15 miles. Subsequent mountain building events have uplifted this ancient core of metamorphic rock while erosion has worn away any overlying sedimentary rocks, exposing the common schist and gneiss we see today.

Igneous Rocks in Front Range

There were two main pulses of igneous **intrusions** into the metamorphic rock complex. The oldest is dated at ~1.7 billion years old. In the Front Range it is known as the Boulder Creek **granodiorite**, named after its type area west of Boulder, Colorado. Between Loveland and Fort Collins, exposures of similar aged intrusions occur at Horsetooth Mountain, Arthur's Rock in Lory State Park, and Bobcat Ridge. Chemically similar varieties of the granodiorite are found at Horsetooth Mountain and Bobcat Ridge and are technically classified as **trondhjemite** and **tonalite**, repectively. And at the iconic Horsetooth Mountain and Arthur's Rock, the minerals composing those intrusions are especially coarse grained and described as **pegmatites**.

A second pulse of igneous activity occurred about 1.4 billion years ago. Compositionally these igneous rocks are classified as **granites**. Along the Front Range there are three separately named but geologically contemporaneous intrusions. The Sherman Granite, well exposed at Eagle's Nest Open Space; the Longs Peak-St. Vrain Batholith, well exposed in Estes Park and Rocky Mountain National Park; and the Log Cabin Batholith, well exposed on trails in the Red Feather Lakes area.

Hard Rock Mining

In 1859, gold was first discovered in Colorado within sediments of Clear Creek at Idaho Springs and nearly simultaneously at Gold Hill just west of Boulder as an in-place vein deposit. The discoveries set off a gold rush that brought tens of thousands of prospectors to the state. Unfortunately, as with all booms, it was relatively short-lived. By 1867 the easy deposits were found and what remained required more capital and technical resources than most fortune-seekers had at their disposal.

The discoveries of gold, silver, copper, and other precious metals mostly fall within a well-defined linear band that stretches from Durango to Boulder. And although Precambrian igneous and metamorphic rocks served as host rocks, the mineralized deposits were emplaced during the **Laramide Orogeny** starting about 70 million years ago. As it turns out, **hydrothermal fluids** rich in minerals and associated igneous activity were focused along a crustal zone of weakness that was established during the collision that formed Colorado 1.8 billion years ago. This zone of weakness defines the Colorado Mineral Belt. The Front Range foothills and mountains of Northern Colorado fall outside this sweet spot of mineralization. Although small mineralized veins have been found near Rustic, Colorado in the Poudre Canyon, they are poorly enriched in gold and uneconomic.

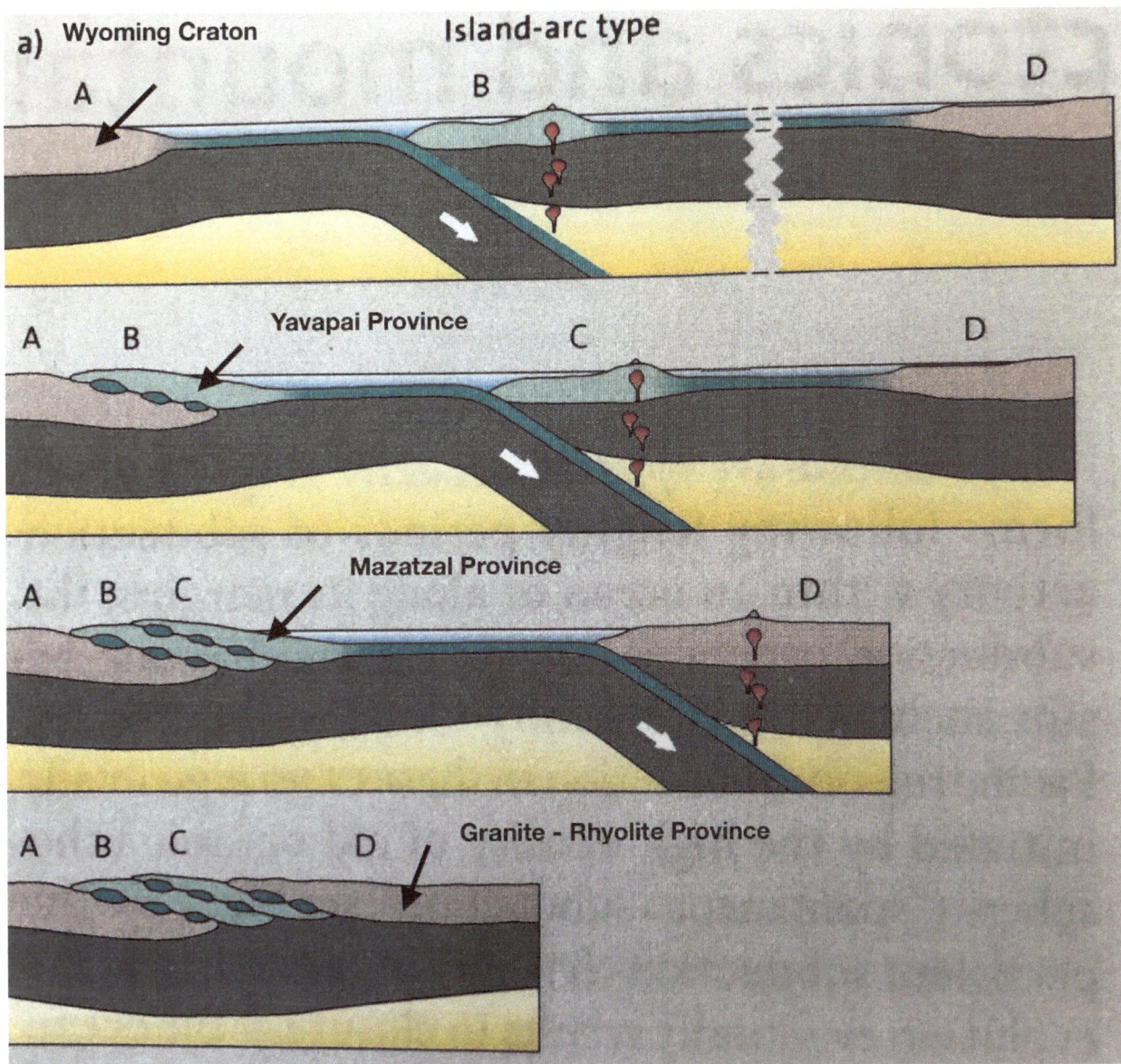

Fig. 10 Plate Tectonic Model

As the conveyer-belt motion of an ancient subducting plate brought the Wyoming **craton** into collisional contact with an offshore **volcanic arc**, the island's volcanic rocks and intervening marine sedimentary rocks were thrust over older rocks of the Wyoming craton along a **shear zone** that dips about 55 degrees to the southeast. The *accreted terrane,* known as the Yavapai Province, not only forms the basement rocks of Colorado, but extends across several states. The process repeated, progressively adding younger accreted terranes to the growing North American continent. Each collision was a mountain building event. The metamorphic gneisses, schists, and many of the igneous rocks we see in our natural areas are the vestigial roots of the ancient Yavapai mountain belt. The Vishnu Schist of the Grand Canyon and Painted Rock in Black Canyon of the Gunnison are other noted locations where the accreted terrain of the Yavapai Province is exposed.
(Modified from Plate Tectonics, Springer-Verlag 2011)

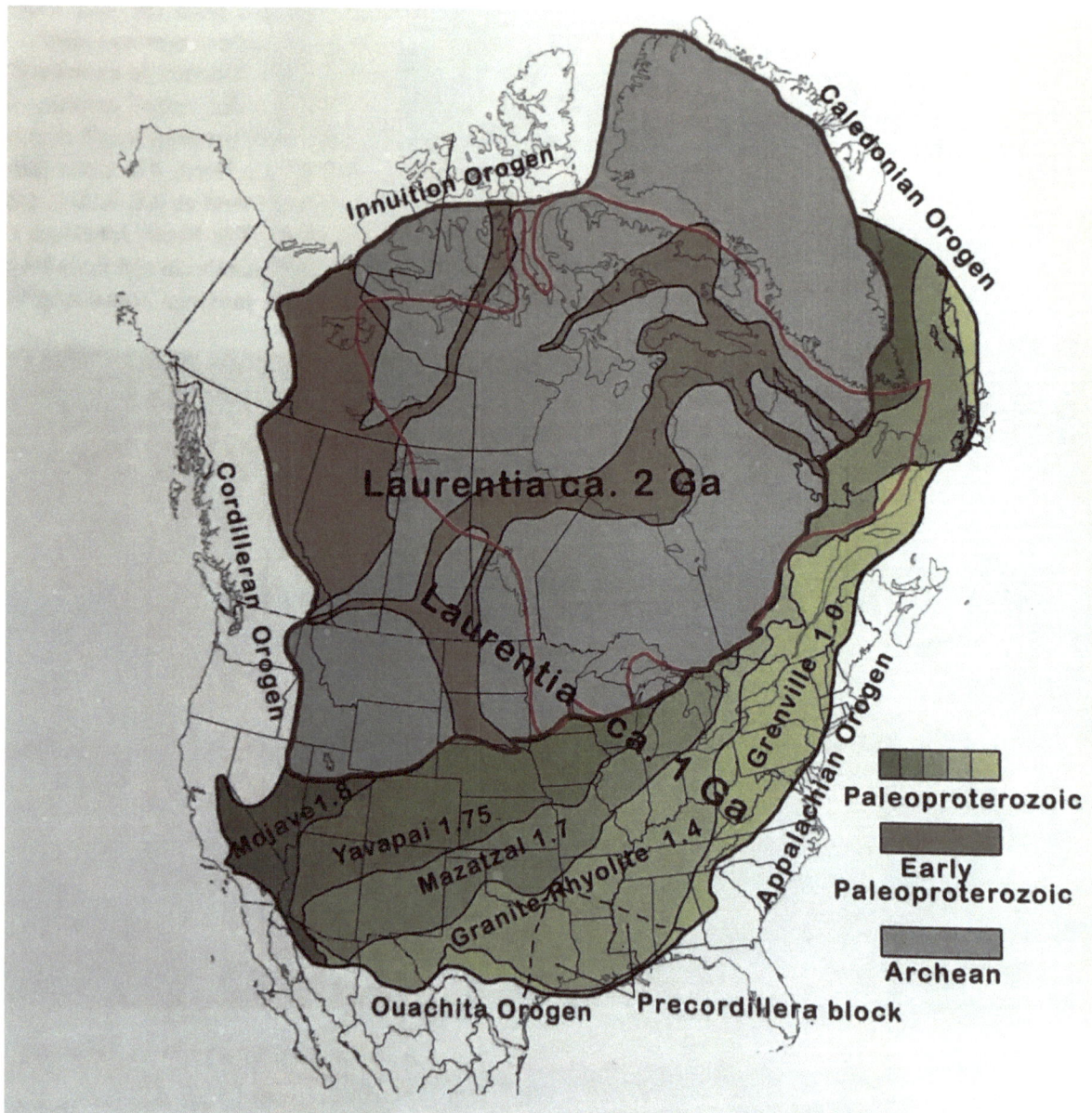

Fig. 11 Stitching North America Together

The size and shape of the North American continent has changed over time through the addition of multiple accreted terranes. The Yavapai terrain, which forms the basement rocks of Colorado, was accreted to the southern margin of an older continent known as Laurentia about 1.75 billion years ago.
(Ancient Landscapes of Western North America, Springer-Verlag 2018)

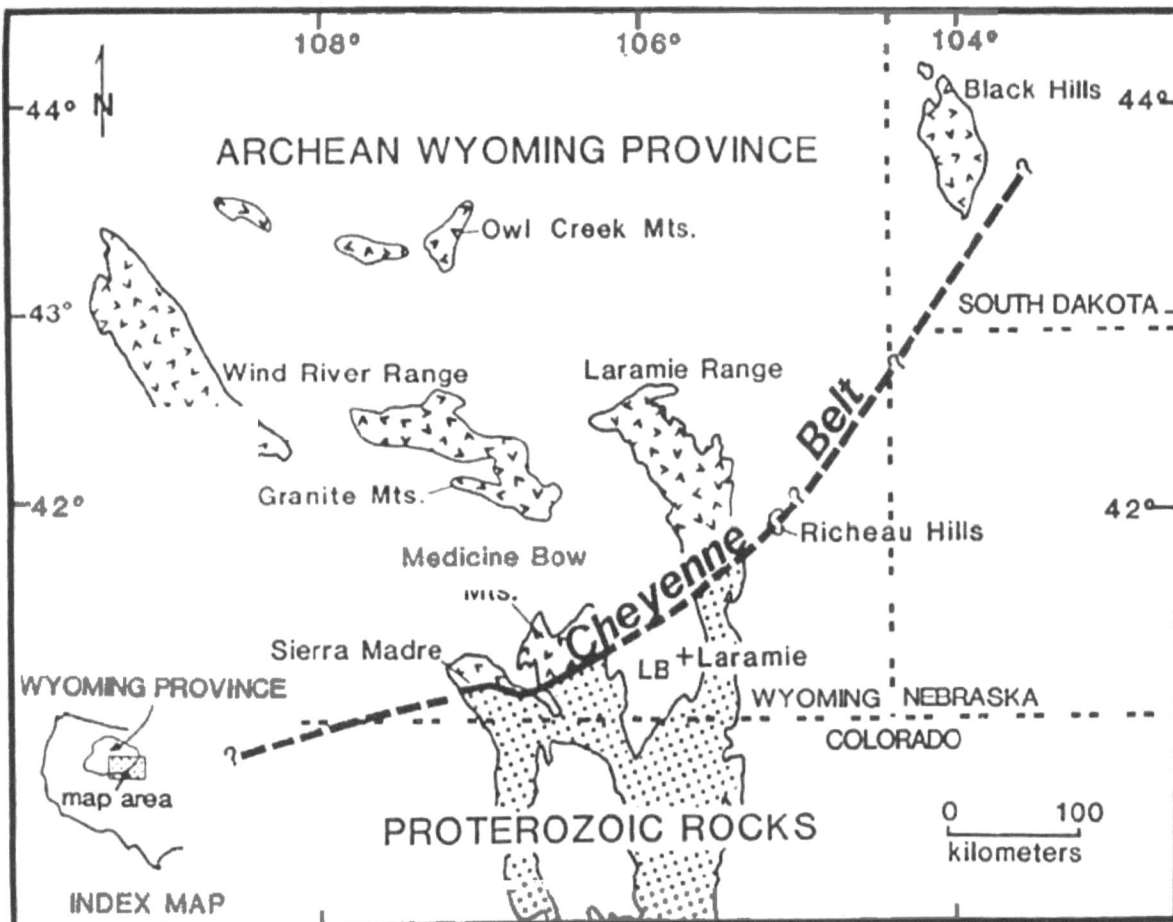

Fig. 12 Suturing Colorado to the Wyoming Craton

At the same time the *Theory of Plate Tectonics* was being developed and accepted by most of the geologic community in the 1960's, geologists recognized a significant boundary within the Snowy Range of the Medicine Bow Mountains in southern Wyoming. The boundary is structurally marked by steeply dipping shear zones that separate older rocks dated greater than 2 billion years old from rocks south of the boundary dated at 1.8 billion years or younger. This boundary, here referred to as the Cheyenne Belt, is now recognized as a well-documented example of a **suture zone,** in this case marking the line along which the Yavapai province became accreted to the Wyoming craton. (Houston RS et al., 1989)

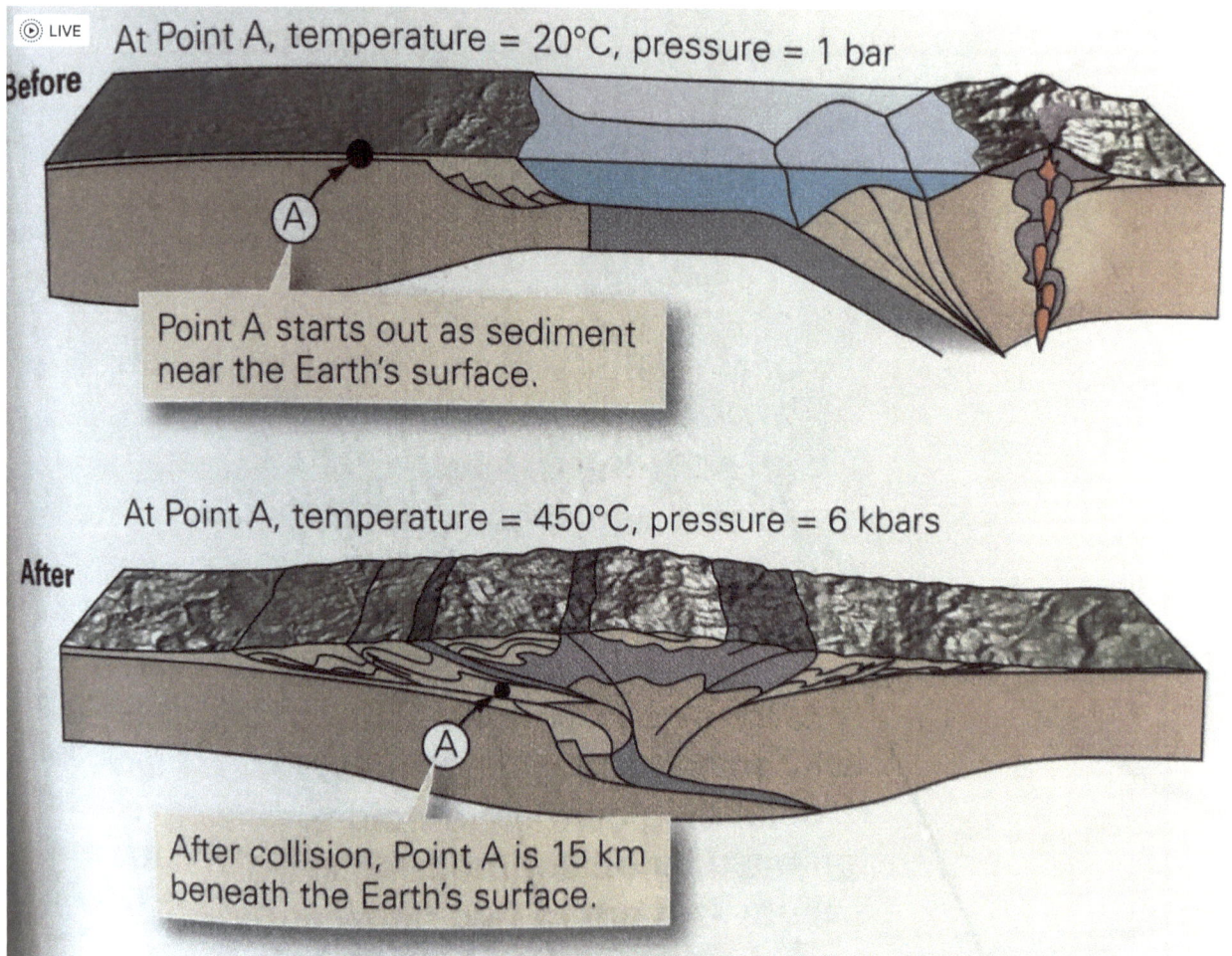

At Point A, temperature = 20°C, pressure = 1 bar

Before

A

Point A starts out as sediment near the Earth's surface.

At Point A, temperature = 450°C, pressure = 6 kbars

After

A

After collision, Point A is 15 km beneath the Earth's surface.

Fig. 13 Metamorphic Environment

When continents collide a belt of metamorphic rocks is created due to deep burial and transformation of original sedimentary or igneous rocks. As thrust sheets stack on top of one another, not only are mountains formed, but the weight of the sheets depresses the **crust**. Rocks that were originally near the surface can sink to depths of 6 miles or more where pressures and temperatures change the original rocks into schists and gneiss, the most common metamorphic rocks in the northern Front Range foothills.
(Earth—Portrait of a Planet, Fifth Edition, Norton 2015)

CLASSIFICATION OF FOLIATED METAMORPHIC ROCKS

ROCK TYPE		DESCRIPTION	INDEX MINERALS	ORIGINAL ROCK
SLATE	INCREASING METAMORPHISM ↓	VERY FINE GRAINED, SMOOTH DULL SURFACE, SPLITS ALONG CLEAVAGE	INCREASING METAMORPHISM → CHLORITE	SHALE, MUDSTONE or SILTSTONE
PHYLLITE		FINE GRAINED, SILKY SHEEN BREAKS ALONG WAVY SURFACES	BIOTITE	SHALE, MUDSTONE or SILTSTONE
SCHIST		MEDIUM GRAINED, PARALLEL ALIGNED MICA MINERALS. SCALY FOLIATION	GARNET	SHALE, MUDSTONE or SILTSTONE
GNEISS		MEDIUM TO COARSE GRAINED ALTERNATING DARK & LIGHT BANDS	STAUROLITE ANDALUSITE SILLIMANITE	SHALE, GRANITE or VOLCANIC ROCKS
MIGMATITE		COARSE GRAINED, GNEISSIC TEXTURE IGNEOUS COMPONENT DUE TO MELTING		SHALE, GRANITE or VOLCANIC ROCKS

Staurolite Schist—
Bobcat Ridge

Sillimanite Schist—Gateway

Hornblende Gneiss—Lory SP

Migmatite—Gateway

Fig. 14 The progression of metamorphism with increasing heat and pressure is marked by changing rock type and the presence of different **index minerals**. Schist and gneiss are the most common metamorphic rocks in the Northern Front Range foothills.

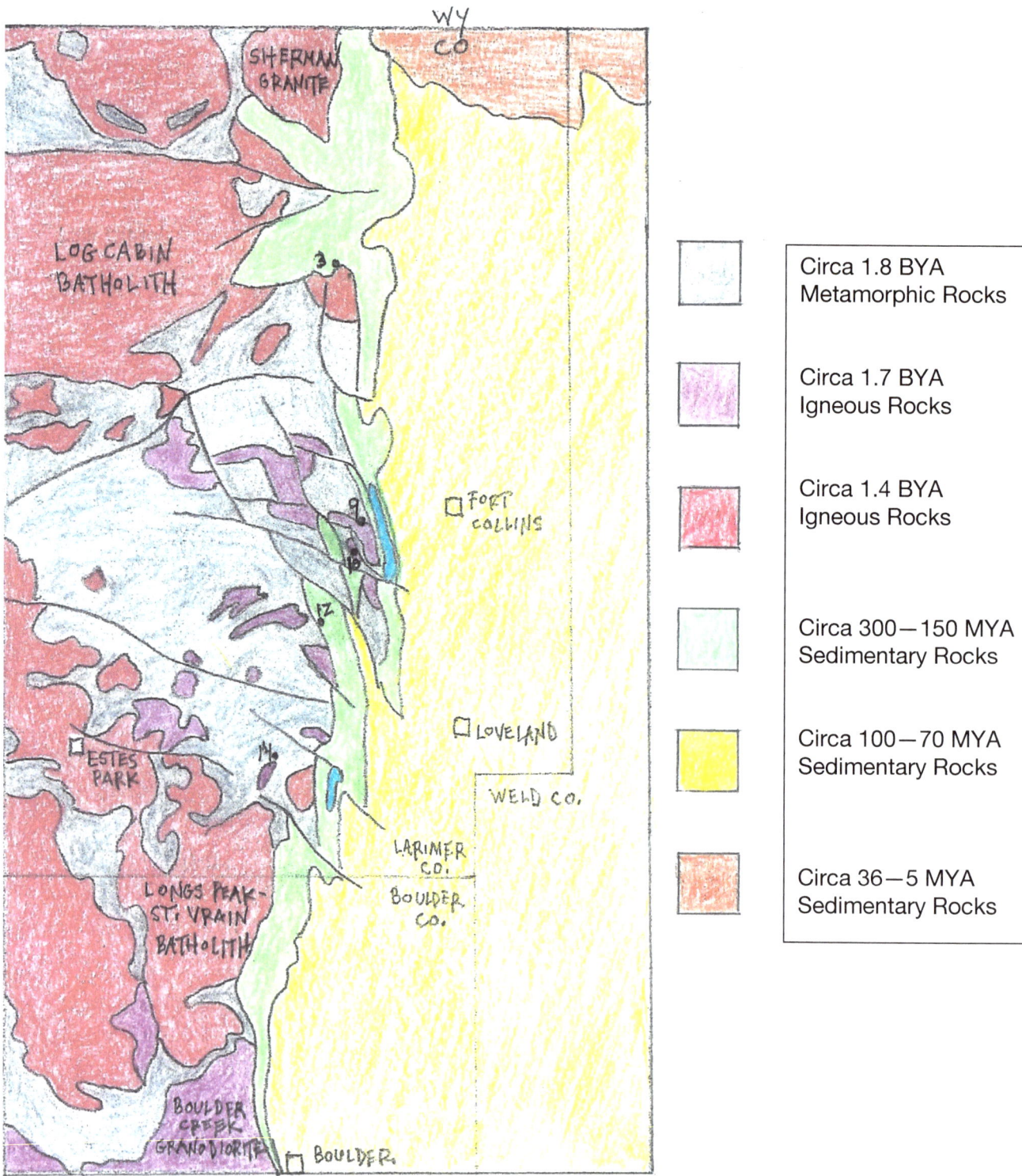

Fig. 15 Regional distribution of basement rocks in the northern Front Range. In the area west of Fort Collins and Loveland, the igneous intrusions are relatively small, and are the same age as the Boulder Creek Granodiorite intrusion west of Boulder. The best natural areas for inspecting the igneous rocks are Eagle's Nest (3), Lory SP (9), Horsetooth Mountain (10), Bobcat Ridge (12), and Ramsay - Shockey (14). (Modified from Workman JB et al., 2018)

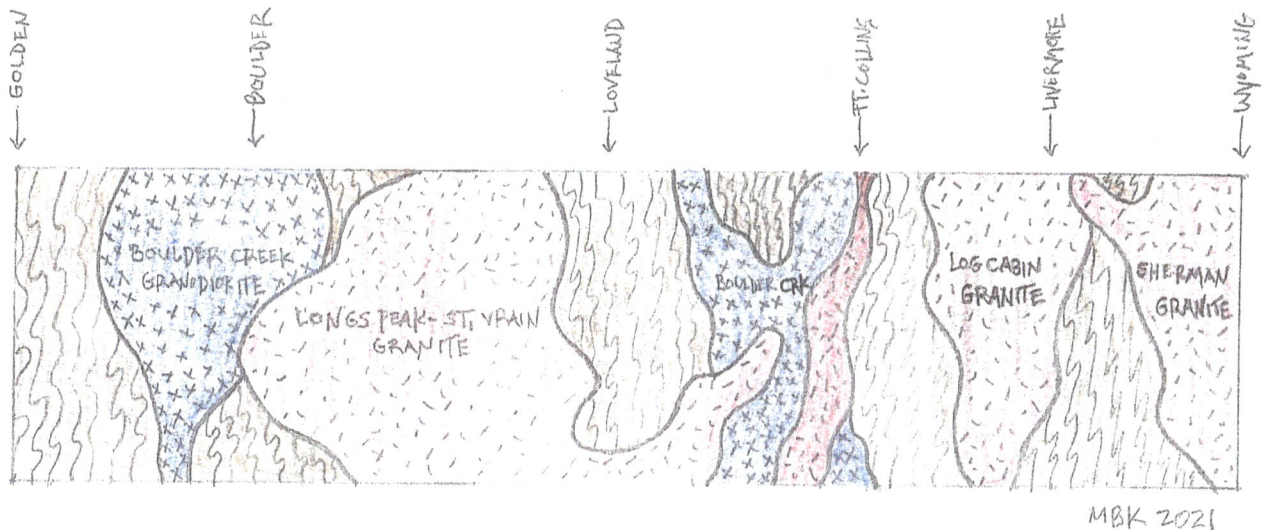

MBK 2021

	Circa 1.8 billion year old metamorphic rocks
	Circa 1.7 billion year old granodiorite, tonalite & trondhjemite
	Circa 1.4 billion year old granite

Fig. 16 Schematic cross section of basement rocks in the Front Range foothills. The names of the igneous bodies are labelled. The 1.7 by old igneous intrusions, often referred to as TTG's for tonalite, trondhjemite and granodiorite, are chemically related rocks that have a higher percentage of feldspars rich in sodium and calcium, compared to the feldspar content of granite. The TTG's were most likely sourced from subducted oceanic crust at the convergent plate boundary between the Wyoming craton and the volcanic island arc.

The granodiorite is exposed in Lory State Park, while the tonalite can be seen at Bobcat Ridge, and the trondhjemite at Horsetooth Mountain and Ramsay-Shockey.

The tectonic setting of the 1.4 BYA granite intrusions is a mystery to geologists. Referred to as anorogenic, or A-type granites, they were emplaced within the old mountain belt produced by continental collision, but several hundred million years after the event.

Pegmatite, which is a textural reference to especially coarse-grained granitic rocks, is prevalent at both Lory State Park and Horsetooth Mountain. Geologists have been unable to determine a date, but suspect the pegmatite was intruded during both the the 1.7 and 1.4 BY igneous episodes.

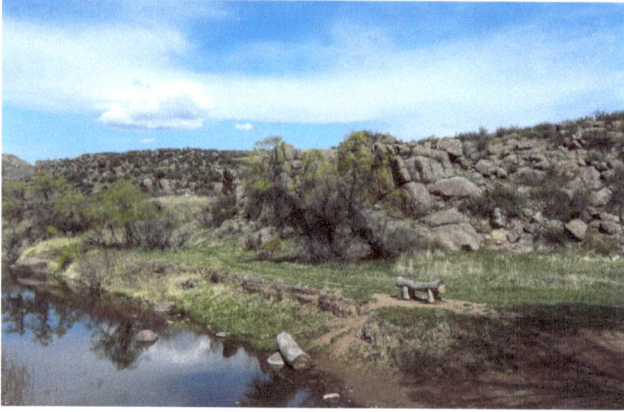

Fig.17a Outcrop of Sherman Granite along the bank of the North Fork of the Cache la Poudre River at Eagle's Nest Open Space.

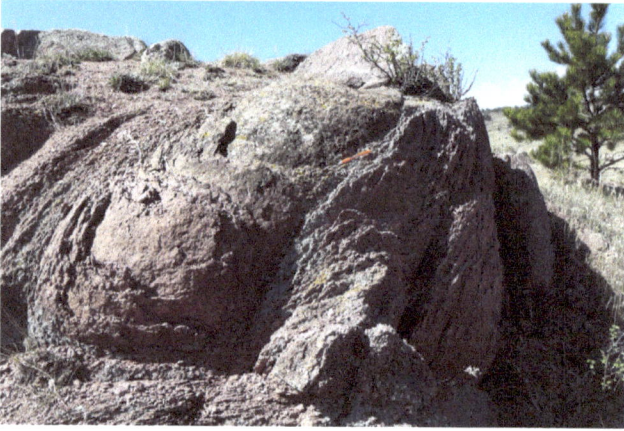

Fig. 17b Outcrop of trondhjemite at Ramsay—Shockey Open Space.

Fig. 17c Close-up of large quartz and feldspar minerals in pegmatite at Lory State Park.

Fig. 17d Contact of pink pegmatite with gray-colored Boulder Creek Granodiorite. Lory State Park.

22

Fig. 17e Horsetooth Mountain in the upper right along with all the exposed rocks in the photo are composed of granite pegmatite.

Fig. 17f Eagle's Nest Rock on the left and the adjacent exposed rocks to the right are slightly younger granites that intruded into the larger body of Sherman Granite at Eagle's Nest Open Space.

Fig. 17g Aerial view (looking north) at Mahoney Park in Bobcat Ridge NA. The light-colored rocks both within and bordering the meadow are composed of tonalite.

FOUNTAIN FORMATION

AGE: 290—310 million years ago (Late Pennsylvanian—Early Permian)

THICKNESS: 800 feet

DEPOSITIONAL ENVIRONMENT: Alluvial fan and braided stream sediments shed off the
Ancestral Rockies, shallow marine

LITHOLOGY: Conglomerates and sandstones interbedded with mudstones and paleosols

---

WHERE ARE GOOD EXPOSURES OF THE FOUNTAIN?

Lory State Park:
 South Valley Trail Loop - Outcrops of cross-bedded conglomerate and sandstone along
 the east segment of the trail
 Shoreline Trail - Contact with overlying Ingleside Fm is exposed near crest of trail. Ingleside
 has an orange color while the Fountain is more maroon colored.
Bobcat Ridge:
 Valley Loop Trail - Exposures of Fountain conglomerate near its erosional edge on west
 segment of loop—bordering exposures of schist.
Ramsay-Shockey:
 Besant Point Trail - Exposure of the Great Unconformity. Fountain overlying Precambrian
 schist.
Watson Lake:
 The maroon colored Fountain Fm forms the slope beneath the cliff-forming Ingleside Fm.

HOW DO I IDENTIFY THE FOUNTAIN?

- Slope and valley former
- Maroon colored with some interbedded grayish-white lenses
- Dominantly a conglomerate and sandstone on most trails
- Often poorly cemented and easily crumbles
- Cross-bedding (see below) evident in many of the outcrop exposures

---

TOPICS OF INTEREST:

The Great Unconformity (GU)

Along the northern Front Range foothills, the 300 million year old Fountain Formation lies
directly on a variety of schists, gneiss and granitic rocks that comprise the 1.8-1.4 billion year
old basement rocks of Colorado. This large gap in the rock record is known as the **Great
Unconformity**. The unconformity was first recognized in the Grand Canyon by John Wesley
Powell during his 1869 Colorado River expedition. Now known to have global extent, the
processes responsible for the formation of the Great Unconformity remain unresolved. Two
popular theories are (1) uplift and erosion of the **supercontinent** Rodinia (1000—750 million
years ago) and (2) glacial erosion associated with "Snowball Earth" conditions. Snowball Earth
is a hypothesis based on geologic evidence that glacial ice covered nearly the entire surface of
the earth during a period known as the Cryogenian 720—580 million years ago. Some
geologists have proposed that both events delivered a one-two punch that resulted in the
erosion of three to five kilometers of rock to explain the unconformity.

The Ancestral Rocky Mountains

The origin of the **Ancestral Rocky Mountains** (ARM) is an active arena of competing hypotheses and debate among geologists. The ARM formed roughly 320−280 million years ago, a period when the North American continent was colliding with South America and Africa to create the supercontinent **Pangaea**. Similar to today's Rocky Mountains formed by the Laramide Orogeny 70−40 million years ago (see Part B), the ARM were composed of basement cored uplifts and adjacent basins that were located far inland of the contemporaneous plate margin. Current research suggests that the ARM were formed by transmitted compressive stresses not only from collision between North and South America as delineated by the Ouachita-Marathon thrust that runs through Texas, Oklahoma and Arkansas but also from stresses originating from the western and southwestern plate margin. The **conglomerates** of the Fountain Formation are rocks composed of **alluvial fan, braided stream** and **paleosol** sediments deposited proximal to the ARM front. The individual pebbles and cobbles within the conglomerate mostly reflect erosion of the igneous granitic terrain that formed the core of the Ancestral Rockies.

Cross Beds

The Fountain Formation is full of sedimentary structures known as **cross bedding** (see Fountain rock column figure). Cross beds are remnants of ripples and dunes created by the flow of water or air over loose sediment. Ripples are typically straight crested features that form on silty or sandy sediment. Only the down-current side of the ripples (foresets) usually gets preserved in the rock record as inclined internal layers bounded by horizontal bedding. This type of cross-bedding is called planar. Trough cross-bedding, the other predominant form, results from more intense flow velocities that result in larger scale, sinuous crested dunes. The observed cross bedding pattern looks like the curved layers of an onion that has been cut in half. Both types of cross bedding are observed in the Fountain.

Red Beds

In the Front Range the geologic formations of Pennsylvanian/Permian age from the Fountain up through the Lykins, are frequently referred to as "**red beds**" owing to their noted color. Composed of mudstones, siltstones, sandstone, and conglomerates, these rocks acquired their red pigmentation when oxygenated ground water came into contact with iron-bearing minerals. The oxidized iron in the form of the mineral Hematite (Fe_2O_3), precipitated as surface coatings around the grains of sedimentary rock. Some lenses or beds within the Fountain are noted by their grayish white color against the more typical maroon coloring of most of the formation. These beds are thought to have been bleached by later migration of reducing petroleum bearing waters from the **Denver Basin** immediately east of the Front Range.

Scenic Geology

In the area of the northern foothills, the Fountain forms a west-facing slope beneath the Ingleside hogback, but erodes to form the adjacent valley floor when not protected by the overlying hard sandstone of the Ingleside. Swiping your hand across a Fountain outcrop typically reveals how poorly cemented it is as the constituent grains easily dislodge from the rock. Further south, the Fountain is more tightly cemented. From Boulder to Colorado Springs, the Fountain has been more steeply tilted and often forms scenic monolithic landscapes such as at the Flatirons in Boulder, Red Rocks, Garden of the Gods, and Roxborough State Park. And even the well-known Maroon Bells in Aspen are Fountain equivalent rocks that were deposited on the western slope. This difference in **cementation** leads to **differential erosion** that can in one case create valleys, and in another, steeply dipping **flatirons**.

SCHEMATIC ROCK COLUMN

FOUNTAIN FORMATION

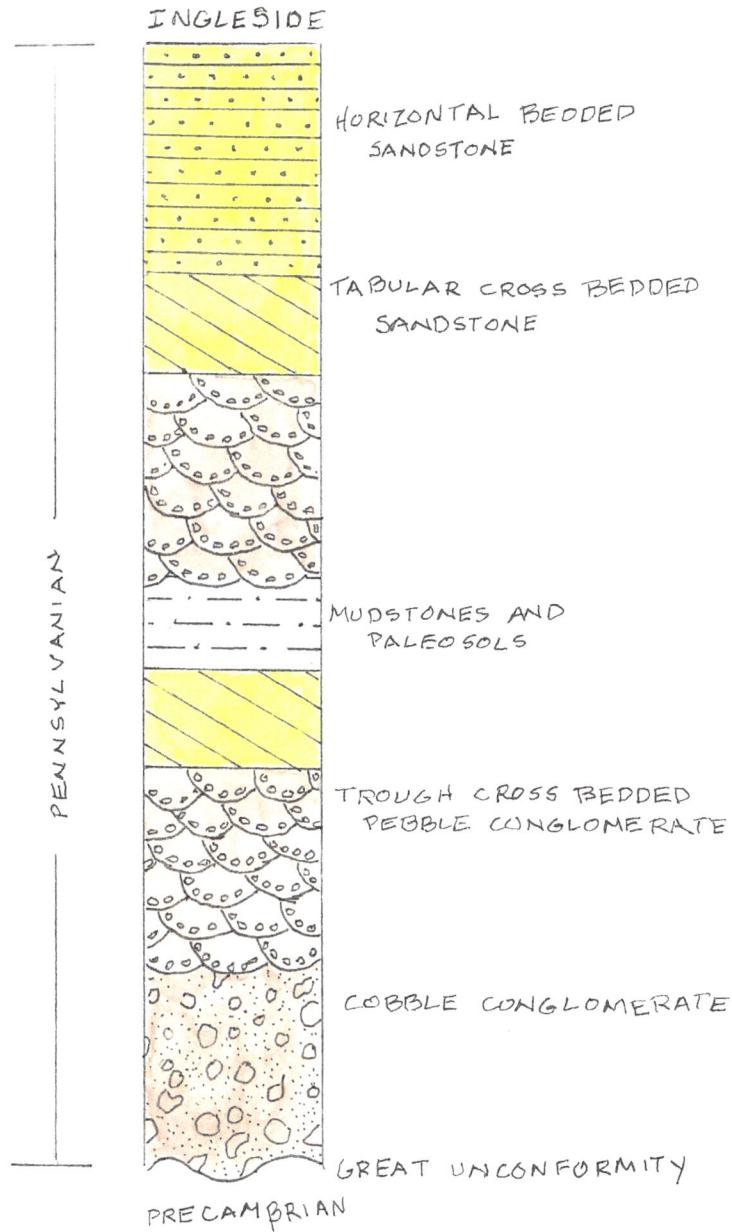

INGLESIDE

HORIZONTAL BEDDED
SANDSTONE

TABULAR CROSS BEDDED
SANDSTONE

MUDSTONES AND
PALEOSOLS

TROUGH CROSS BEDDED
PEBBLE CONGLOMERATE

COBBLE CONGLOMERATE

GREAT UNCONFORMITY

PENNSYLVANIAN

PRECAMBRIAN

Fig. 18 The Fountain Formation reflects deposition on alluvial fans, braided stream floodplains and adjacent shallow marine environments.

Fig. 19 The Great Unconformity is wonderfully exposed along the north bank of the Charles Hansen Feeder Canal just north of Flatiron Reservoir and CR18E. Here the reddish colored Fountain Fm overlays dark colored schist.

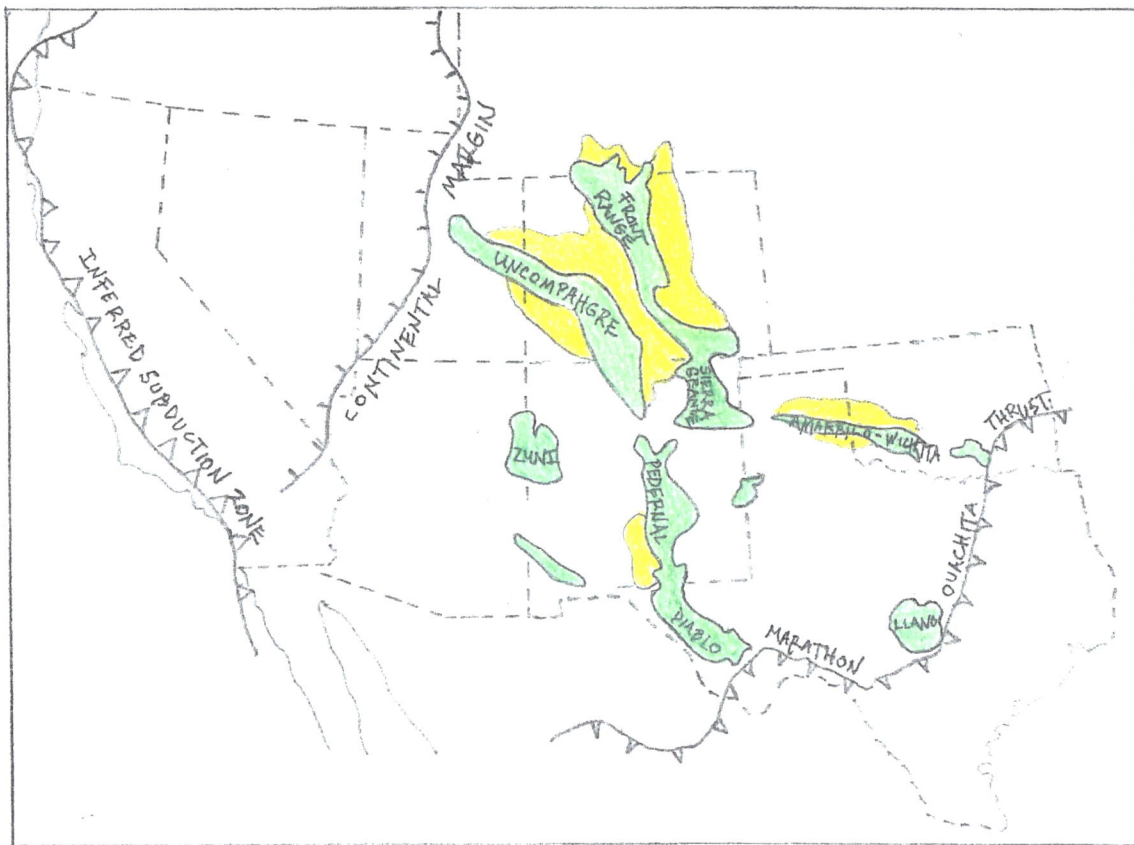

Fig. 20 Basement-cored uplifts of the Ancestral Rocky Mountains are noted in green. The two principle uplifts in Colorado were the ancestral Front Range and Uncompahgre. The areas colored in yellow represent mostly coarse conglomerates and sandstones eroded from the uplifts. The cause of the uplifts remains unsettled. Their orientation and distance from the collisional plate margin marked by the Marathon-Ouachita Thrust is problematic. Many geologists believe their orientation reflects inherited basement weaknesses that were developed in the Precambrian. Throughout the Pennsylvanian and Permian Periods, these mountain uplifts stood as islands surrounded by an interior sea during a period of high sea-level known as the Absaroka transgression. (Modified from Baldridge, 2004)

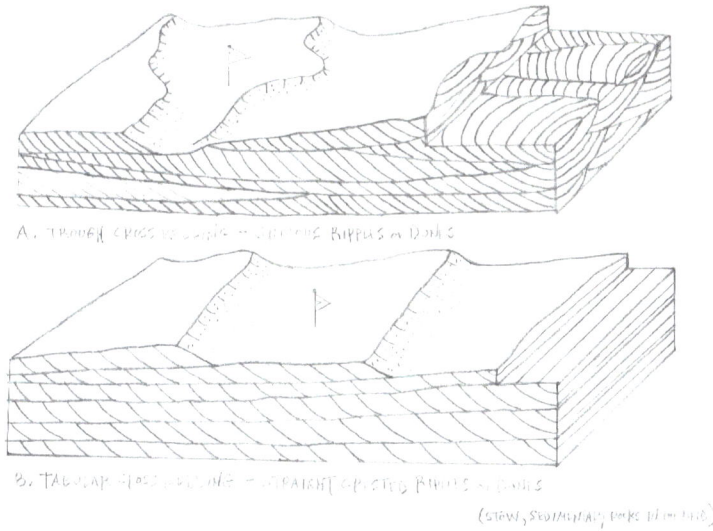

Fig. 21 Trough and tabular cross-bedding often seen in the Fountain, Ingleside and Lyons formations. (Modified from Sedimentary Rocks in the Field, 2011)

Fig. 22a Fountain outcrop along the South Valley Loop Trail in Lory SP. Here a channel is cutting down into older cross-bedded channel deposits.

Fig. 22b The South Valley Loop Trail parallels the Ingleside hogback ridge. About two-thirds of the way down the cliff face is the contact between the orange-colored Ingleside and the maroon-colored Fountain. The vegetated slope below is also Fountain Fm rocks.

Fig. 22c Close up view of the contact between the Fountain and Ingleside formations. The Fountain is the recessed pock-marked unit with gray splotches. In contrast to the conglomerates in the lower part of the Fountain, here it is made up of finer-grained beach sand.

Fig. 22d Fountain Fm outcrop off Besant Point Trail at Ramsay-Shockey OS. Here a channel filled with coarse cobble conglomerate cuts into sandstone.

Fig. 22e Chimney Hollow Dam under construction. The Fountain Fm comprises the entire vegetated slope below the Ingleside ridge line. Overburden and weathered Fountain Fm rock is being removed to provide a clean and competent surface for the Dam's east wall. The tunnel on the lower left through the Fountain will be used for reservoir inlet/outlet.

INGLESIDE FORMATION

AGE: 285—290 million years ago (Early Permian)

THICKNESS: 150—240 feet

DEPOSITIONAL ENVIRONMENT: Shallow marine and coastal dunes

LITHOLOGY: Fine grained quartz sandstone and limestone

WHERE ARE GOOD EXPOSURES OF THE INGLESIDE?

Devil's Backbone:
   Blue Sky Trail - Outcrops along both the Hunter and Laughing Horse Loops.
   Rimrock Trail - Exposures on the short loop and at the top of the ridge.

Lory State Park:
   Shoreline Trail - Outcrops along the spur trail heading south from the top of the
                    ridge all the way to Quarry Cove overlook.

Red Mountain:
   Bent Rock Trail - Exposed along the walls of Sand Creek Canyon cutting through
                    Bent Rock. The Ingleside is mostly limestone here.

Watson Lake (Bellvue Dome): Ingleside forms the vertical cliff face.

HOW DO I IDENTIFY THE INGLESIDE?

- Westernmost hogback in the foothills
- Red-orange color
- Fine grained quartzose sandstone (limestone dominant at Red Mountain OS)
- Frequently cross-bedded
- Frequently weathers into thin-bedded flagstone slabs

TOPICS OF INTEREST:

Historic Quarrying

Starting at Watson Lake (Bellvue Dome) and north to Wyoming, the Ingleside formation becomes increasingly interbedded with limestone reflecting a more prominent marine depositional setting. In 1906, the Ingleside Limestone Company (a subsidiary of Great Western Sugar Company) opened a quarry at the namesake town located off US 287 between Ted's Place and Owl Canyon. Limestone ($CaCO_3$) was found to be a key ingredient for purifying beet juice into granulated sugar. The limestone eventually played out at this location and the quarry was abandoned in the 1930s when new quarry operations moved north to Owl Canyon. The quarries are now active under new ownership, but with the demise of the sugar beet industry, Ingleside limestone and sandstone are now being used for various other construction and manufacturing purposes. Both quarries are easily seen from US 287.

Additionally, from Bellvue south to Orchard Cove on the west side of Horsetooth Reservoir, the Ingleside sandstone was quarried in large blocks starting in the late 19th century. The Union Pacific Railroad had built rail lines to Bellvue and south to the historic town of Stout where UPR owned its own quarries. The railroad greatly expanded markets for the building stone beyond Colorado. All the Horsetooth Ingleside quarries ceased operations in the 1930s prior to the filling of Horsetooth reservoir.

Bird Nests

The well cemented quartz **sandstone** of the Ingleside Formation is highly resistant to erosion and forms the westernmost hogback within the foothills. The western face, or **escarpment** side of the hogback typically forms a near vertical cliff with some ledges created by differential erosion. Vertical relief from the top of the cliff to the valley bottom below can be 200 feet or more. Raptors find the geology perfectly suited for building protected nests within the Ingleside cliff face. At Watson Lake SWA, golden eagles can often be seen soaring above their nests at Bellvue Dome. And on the Rimrock Trail at Devil's Backbone Open Space, falcons have a nest built into the Ingleside cliff face and are often seen gliding in the skies above. Also look for swallow nests built on the underside of ledges that protrude from the Ingleside cliff face.

Bellvue Dome

The spectacular setting of Bellvue Dome, the name given to the giant rock edifice jutting up over the Cache La Poudre River and overlooking Watson Lake is a special place to visit. At first glance one might suspect this is just another tilted east-dipping hogback. But, following the ridgeline, the rocks can be seen to dip both northward and southward on the respective sides of the uplift. And on the west side of Watson Lake, there is a much smaller outcropping of rocks that appear to dip nearly vertically but slightly westward into the earth. Actual measured dips confirm that the rocks all dip away from a projected high position now located over Watson Lake. Erosion removed the crest of the structure, so you have to use your imagination to connect the east and west dipping rocks on either side of Watson Lake (Fig. 24). Geologists have identified a thrust fault (unsurprisingly named the Bellvue Fault) that runs along the west side of the lake; it was active during the Laramide Orogeny and created this type of **fold** in the rocks known as an **anticline**. The term 'dome' is a bit of a misnomer as it implies an unfaulted symmetrical arching of the rocks. The Bellvue Dome and similar anticlines exposed at Red Mountain Open Space (see Fig. 29) and Devil's Backbone Open Space (see Fig. 57) are technically referred to as fault-propagation folds because the rocks were bent over the tip of an upward propagating thrust fault.

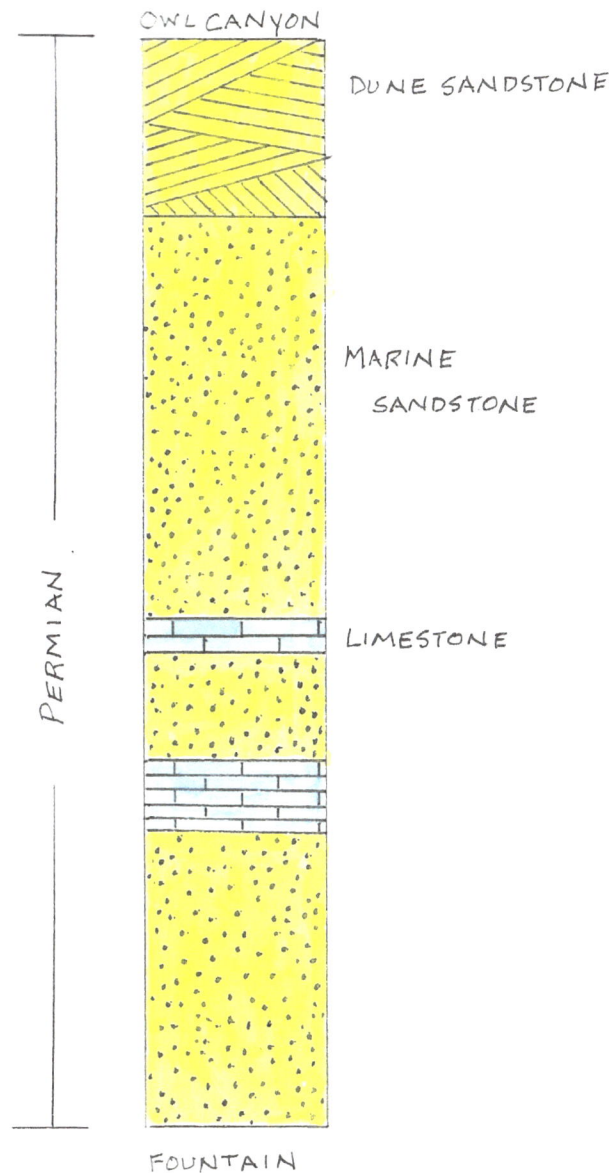

SCHEMATIC ROCK COLUMN
INGLESIDE FORMATION

OWL CANYON

DUNE SANDSTONE

MARINE SANDSTONE

LIMESTONE

PERMIAN

FOUNTAIN

Fig. 23 The Ingleside Formation is composed of rock layers reflecting deposition as onshore sand dunes as well as shallow marine sandstone and limestone. The thickness and number of limestone layers increases from Bellvue to Red Mountain Open Space.

GEOLOGIC CROSS SECTION
BELLVUE DOME

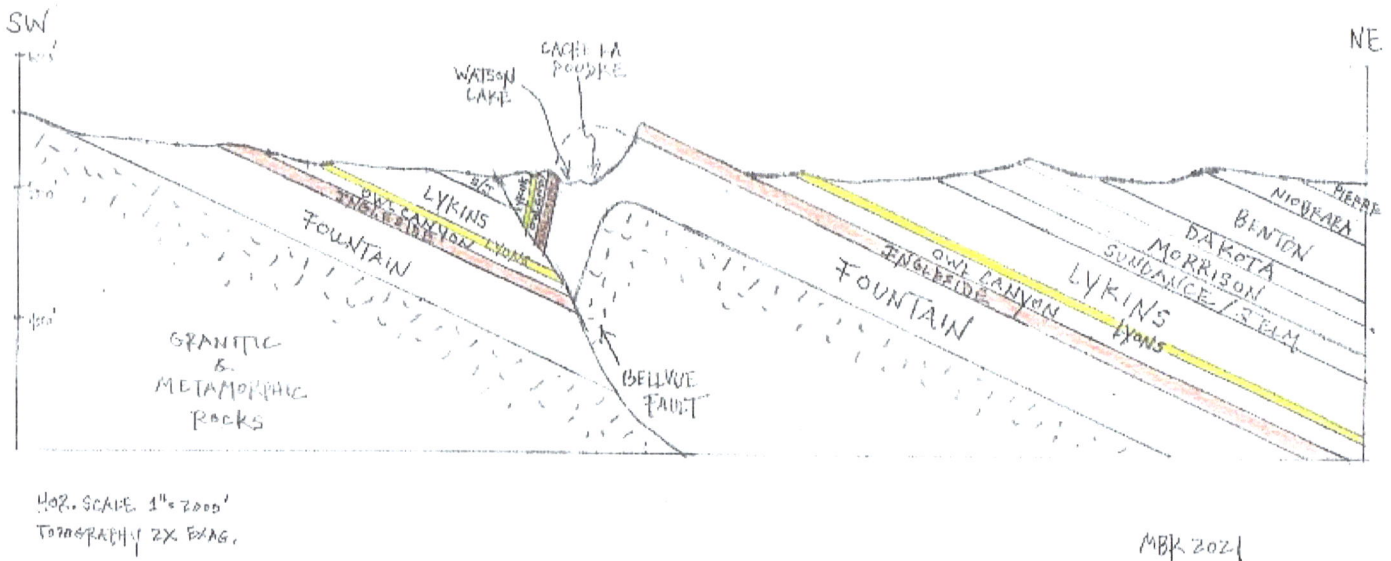

Fig. 24 Bellvue Dome is an asymmetric anticline with steep west dipping beds on its west flank adjacent to the NE dipping Bellvue Fault.

Basement-Involved, Second-Order Anticlinal Structures

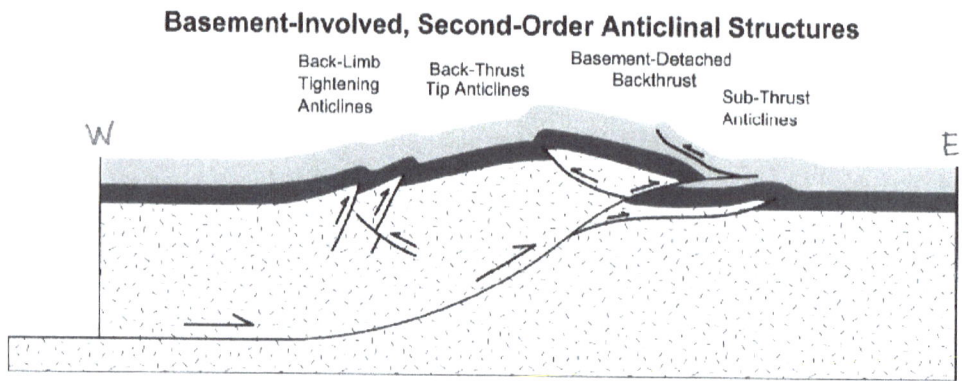

Fig. 25 Structural model illustrating how several prominent anticlines have been formed by shallow back-thrust faults near the east end of the master thrust fault. Examples include the Milner Mountain Anticline (Devil's Backbone), Bellvue Dome (Watson Lake) and the Sand Creek Anticline (Red Mountain OS). (Larson SM, 2009)

Fig. 26a View across Watson Lake to Bellvue Dome. Vegetated slope is Fountain Fm. The cliff face is Ingleside —home to eagle and red-tail hawk nests.

Fig. 26b Beautiful dune cross beds displayed in Ingleside outcrop located on the road into Bobcat Ridge.

Fig. 26c Cliff swallow nests built into the Ingleside Fm. Bobcat Ridge Natural Area.

Fig. 26d Dune cross beds seen off the Shoreline Trail at Lory SP.

Fig. 26e Ingleside quarry just west of US 287 and the small town of Ingleside.

Fig. 26f Cross bedded Ingleside sandstone on the Hunter Loop at Devil's Backbone.

OWL CANYON FORMATION

AGE: 285—278 million years ago (Early Permian)

THICKNESS: 200—350 feet

DEPOSITIONAL ENVIRONMENT: Coastal plain, tidal flats, estuary bay

LITHOLOGY: Mudstone, siltstone, and sandstone

WHERE ARE GOOD EXPOSURES OF THE OWL CANYON?

In Red Mountain Open Space (RMOS), the Lyons Sandstone thins to just a few feet in thickness, or has been completely eroded exposing the Owl Canyon at the surface.

Red Mountain:
    Bent Rock Trail - East side of trail traverses on top of well exposed Owl Canyon with some remnants of white colored Lyons Sandstone present. West side of trail presents a good view of steeply dipping Owl Canyon flatirons forming west flank of Bent Rock.
    K-Lynn Cameron Trail - West half of loop traverses directly on Owl Canyon.

Lory State Park:
    Orchard Cove Trail - When Horsetooth Reservoir is seasonally drawn down Orchard Cove provides an accessible exposure of the Owl Canyon on the south side of the Cove below the Lyons Sandstone ridge.
    Shoreline Trail - Look for thin ribs of exposed Owl Canyon protruding from the valley floor between the Lyons and Ingleside hogbacks.

HOW DO I IDENTIFY THE OWL CANYON?

- Composed of red mudstone, siltstone and fine-grained ripple laminated sandstone
- Typically covered by vegetation
- Forms west-facing slope beneath Lyons hogback ridge, and shallow valley east of Ingleside hogback

---

TOPICS OF INTEREST:

Sand Creek Anticline

One of the most interesting features in Red Mountain Open Space is the exposure of segments of a narrow belt of north-south trending folded rocks that run down the middle of the broad valley named the Big Hole. The folded rocks form a wrinkled pair of a trough shaped **syncline** on the west and an arch shaped **anticline** on the east. The fold belt runs for at least five miles, truncated on the north end by the overlying flat bedded rocks of the Ogallala Formation; while on the south end they appear to die about one mile south of the prominent **butte** known as Table Mountain.

The anticlinal trend is referred to as the Sand Creek Anticline. The topographic high known as Bent Rock is one segment of the anticline. An appreciation of the anticlinal form is best seen from within Sand Creek Canyon which cuts through the anticline. From here a cross-sectional view of the fold is presented. Note the directional change in orientation of the beds from relatively gently east dipping rocks on the east side of the canyon to very steeply west dipping rocks on the west side of the canyon.

Inverted Topography

Rocks that have been folded into an anticline are frequently associated with elevated topography. But weathering and erosion can quickly go to work on elevated topography, beveling off and incising into the top of a fold thereby exposing older rock in the core of the fold. In this way topographical lows or anticlinal valleys can develop on the top of what was once a structurally and topographically high feature. And by similar paradox, rocks folded into trough-shaped synclinal valleys can develop into topographic highs if the rocks within the valley are resistant to erosion. This situation is referred to as **inverted topography**.

At RMOS there are two prominent topographic features that reflect inverted topography. Table Mountain, south of Bent Rock, appears as a free-standing butte rising about 800 feet above the valley but is geologically part of the synclinal fold that parallels Sand Creek Anticline. And on the north side of RMOS, another topographically high feature (unnamed, but north and east of Big Hole Wash Trail) is also part of the same synclinal fold trend. Both of these topographic highs are capped by the Dakota Sandstone. Measured dips across both features indicate that they are both remnants of rocks that have been folded into a trough shaped syncline. A good view of the synclinal fold can be seen looking north from the west side of Bent Rock Trail or from the K-Lynn Cameron Trail where the trough shaped orientation of the rocks is well displayed on the south face of the unnamed topographically high feature.

At Bent Rock, which is a remnant of the anticlinal fold, the cap rock is the Owl Canyon Formation, all the younger rocks, which have been preserved on the two inverted synclinal features discussed above, have been stripped off by erosion.

Superposed Stream

After exiting Haygood Canyon on the northwest side of RMOS, Sand Creek flows south following the Ruby Wash Trail. But rather than continuing to flow south on the west side of the valley where it would join as a tributary to Boxelder Creek, it makes an abrupt eastward turn and cuts a canyon through the Sand Creek Anticline, effectively dissecting the feature known as Bent Rock into north and south halves. As streams normally flow around topographically high features, what would cause Sand Creek to cut across and through Bent Rock?

In 1869, during the start of his famous voyage down the Colorado River, John Wesley Powell observed that the Green River cuts through the eastern side of the Uinta Mountains rather than flowing around it. His theory, which he called **antecedence**, was that the river was older than the mountains and that as the mountains were uplifted, the river kept pace and simply cut down into the uplift. His theory came to be challenged with an alternative known as **superposition**. In this theory, the river is younger than the mountains which were buried by younger flat lying sediments after the mountain building ceased. The drainage patterns that were established on top of this subdued topography were imprinted onto the older mountain topography as the younger flat-lying sediments were ultimately stripped away by erosion. Sand Creek's drainage pattern was likely established on top of the younger flat lying Ogallala Formation which erosion subsequently stripped off. Sand Creek simply cut down and across the older folded rocks created during the Rocky Mountain uplift.

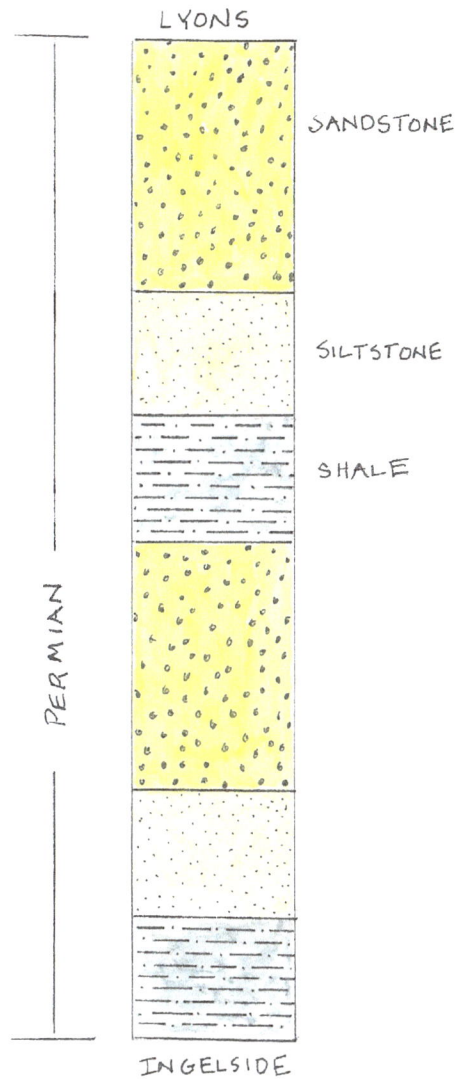

SCHEMATIC ROCK COLUMN
OWL CANYON FORMATION

LYONS

SANDSTONE

SILTSTONE

SHALE

PERMIAN

INGELSIDE

Fig. 27 The reddish-brown Owl Canyon Formation is composed of mudstone, siltstone and fine-grained sandstone deposited on a coastal plain/tidal flat environment and within calm shallow water. Sedimentary structures such as cross-bedding are rare.

39

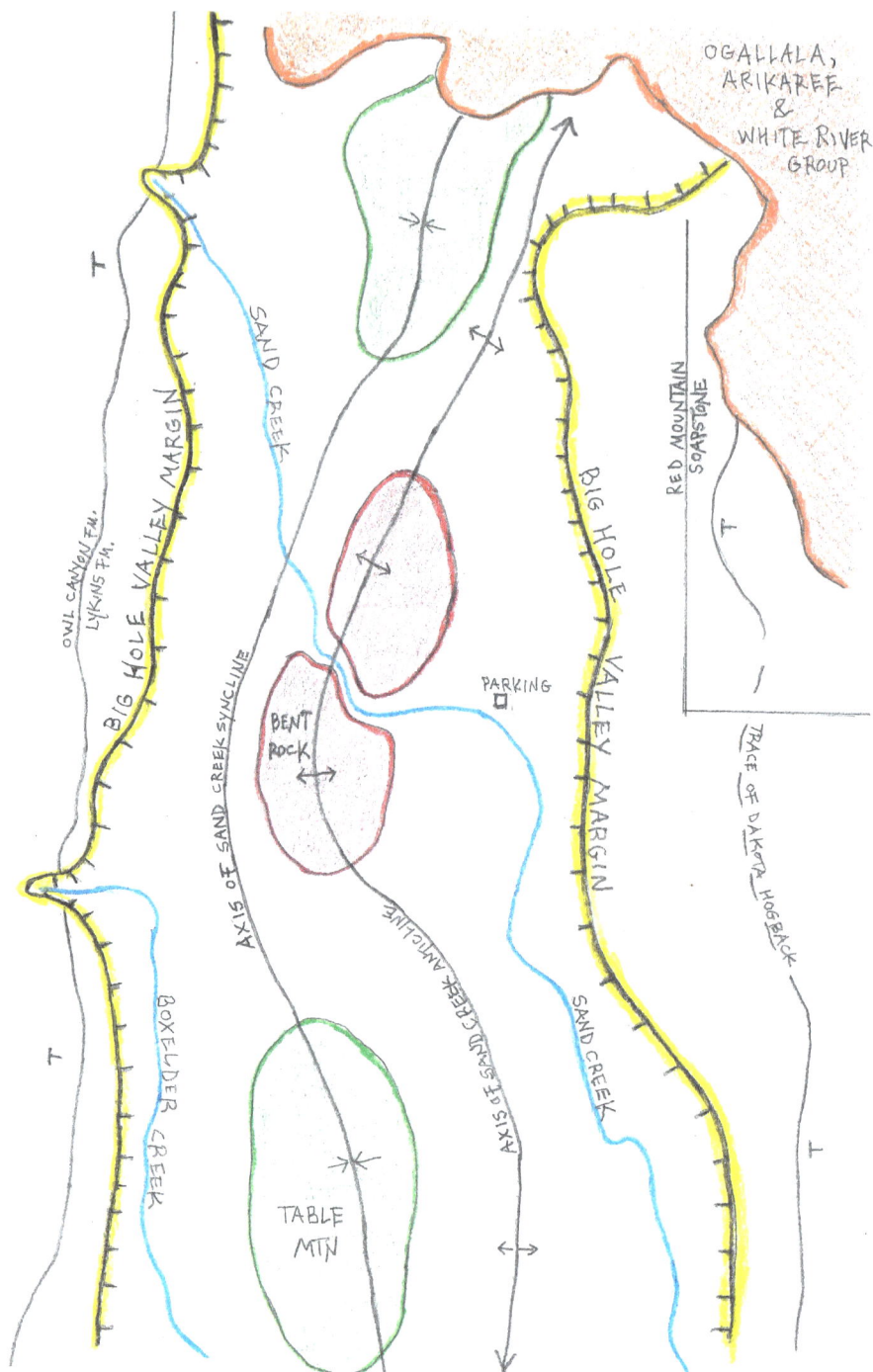

Fig. 28 Sketch map showing the prominent topographic highs within the Big Hole Valley at Red Mountain Open Space. The highs outlined in green are synclinal folds that differential erosion has shaped into high standing buttes. Sand Creek is shown cutting through Bent Rock, a well exposed segment of the Sand Creek anticline.

Fig. 29 (next page) Geologic cross section through Red Mountain and Soapstone Prairie. Sand Creek anticline in Red Mountain likely formed by a basement thrust fault. Soapstone Prairie is underlain by younger Cretaceous rocks and capped by gently dipping Oligocene and Miocene aged rocks.

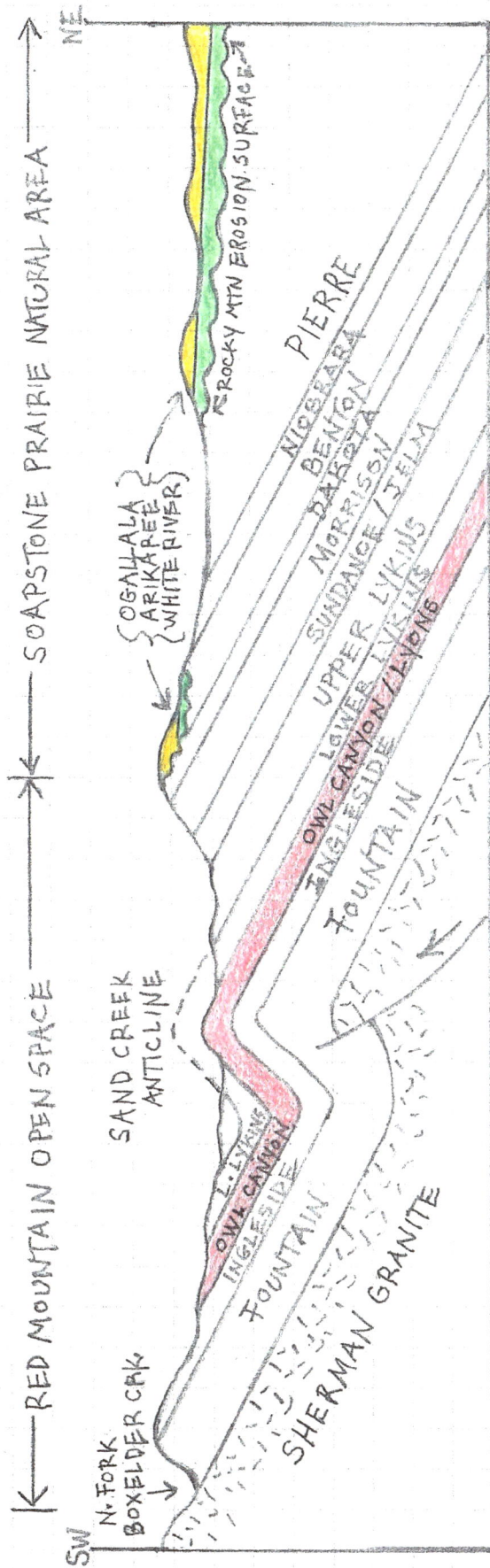

SW — RED MOUNTAIN OPEN SPACE — ‖— SOAPSTONE PRAIRIE NATURAL AREA — NE

N. FORK
BOXELDER CRK

SAND CREEK ANTICLINE

OGALLALA
ARIKAREE
(WHITE RIVER)

← ROCKY MTN EROSION SURFACE →

PIERRE

NIOBRARA

BENTON

DAKOTA

MORRISON

SUNDANCE / JELM

UPPER LYKINS

LOWER LYKINS / LYONS

OWL CANYON

INGLESIDE

FOUNTAIN

L. LYKINS

OWL CANYON

INGLESIDE

FOUNTAIN

SHERMAN GRANITE

41

Fig. 30
Aerial view looking south at the Sand Creek Canyon cut through Bent Rock—part of the Sand Creek anticline extending through Red Mountain. The red rocks are Owl Canyon. The inner canyon walls are white colored limestone of the Ingleside Fm. Note the flatirons formed on the steeply dipping beds of the Owl Canyon.

Fig. 31
Development of a superposed stream downcutting through an anticline. A plausible explanation for why Sand Creek cuts through Bent Rock at Red Mountain Open Space. (Understanding Earth, Fourth Edition, Freeman, 2004)

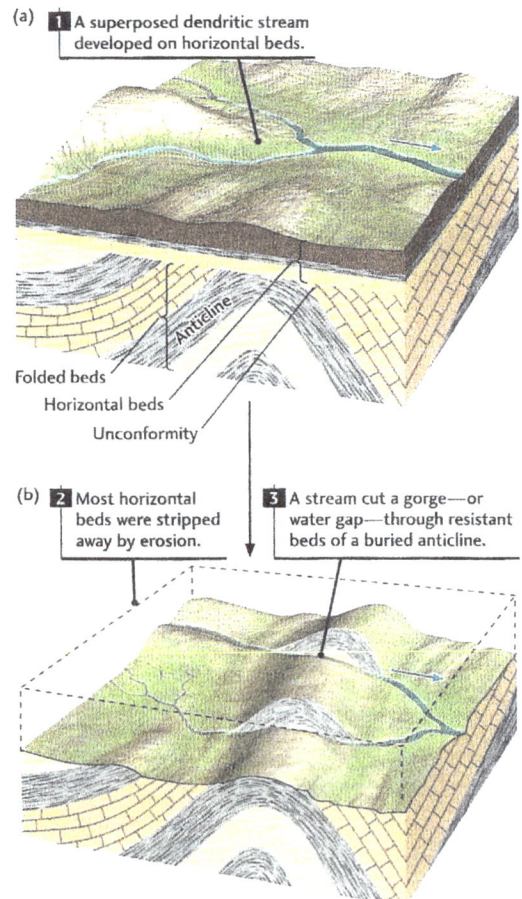

(a) **1** A superposed dendritic stream developed on horizontal beds.

Folded beds
Horizontal beds
Unconformity
Anticline

(b) **2** Most horizontal beds were stripped away by erosion.

3 A stream cut a gorge—or water gap—through resistant beds of a buried anticline.

Fig. 32a Here the reddish-brown Lykins Fm directly overlies the rust-colored Owl Canyon Fm. The Lyons Sandstone is missing due to either erosion or non-deposition.
(Red Mountain Open Space, K-Lynn Cameron Trail).

Fig. 32b Nice exposure of the Owl Canyon Fm on the road into Bobcat Ridge Natural Area.

Fig. 32c View looking east from the top of the Ingleside Fm hogback in Lory SP. The Owl Canyon Fm rocks are the thin beds forming the slope beneath the Lyons Sandstone ridge.

Fig. 32d When Horsetooth Reservoir is drawn down, the red-bedded Owl Canyon Fm is nicely exposed within Orchard Cove. The upper cliff face is Lyons Sandstone.

Fig. 32e Aerial view looking north along plunging Sand Creek Anticline. The anticline is cut in half by Sand Creek Canyon. The southern half is referred to as Bent Rock. The red beds are Owl Canyon Fm rocks. To the north, the topographically inverted Sand Creek Syncline is well exposed—note the bowl shaped attitude of the beds. Further north, flat-laying, post-Laramide rocks cap the High Plains plateau. The Rocky Mountain Erosion Surface separates the younger flat-laying rocks from the folded rocks below.

LYONS SANDSTONE

AGE: 278—274 million years ago (Early *Permian*)

THICKNESS: 30—50 feet (thins to a few feet at Red Mountain Open Space)

DEPOSITIONAL ENVIRONMENT: Desert dune and interdune

LITHOLOGY: Quartz sandstone, siltstone

WHERE ARE GOOD EXPOSURES OF THE LYONS?

Lory State Park:
    Shoreline Trail - Trail cuts through the Lyons hogback before descending to Horsetooth Reservoir. Stepping off the trail it is easy to examine the dune cross-bedding on the west face of the hogback.

Devil's Backbone:
    Rimrock Trail - On the west end, the trail cuts through a water-gap in the Lyons Sandstone hogback. Contact with the underlying Owl Canyon Fm. Is exposed.

Bobcat Ridge:
    Entrance Road - At the turn in to Bobcat Ridge NA (CR32C), adjacent to the active quarry, cross-bedded Lyons Sandstone outcrops on both sides of the road.

HOW DO I IDENTIFY THE LYONS?

- The middle of the three foothills hogbacks (Dakota, Lyons, Ingleside)
- Large scale cross-bedding
- Yellow gray to pale red color
- Composed almost entirely of fine-grained quartz

TOPICS OF INTEREST:

Stone Quarrying

Long before European settlers arrived, artifacts have shown that American Indians of the Archaic period (6500 BC—200 AD) were the first to use the flagstone-like layers of the Lyons Sandstone as a base to mill wild plant seeds. With the settling of the Front Range in the late 19th Century, the rock began to be exploited as a construction resource and quarrying soon scaled to industrial proportions.

In the Lyons area, settler Chester Smead was the first documented to quarry the rock in 1873 for the building of a school and church in Longmont. Later in 1880, Edward S. Lyon, a settler from Connecticut, purchased a 160 acre homestead to pursue farming and ranching. Mr. Lyon soon realized that more profit could be had from the commercial value of the red sandstone outcrops on his property than from cattle and crops. A town sprang up to support his quarrying operation and a schoolhouse was built in 1881. The schoolhouse now serves as the town museum and along with 14 other red sandstone structures in Lyons are on the National Register of Historic Places.

Just west of Fort Collins, the town of Stout was established in the 1870s to serve as a base of Lyons Sandstone quarrying operations that were in commercial competition with the Lyons quarries. By the 1920s challenges to the quarrying industry and the impending development of Horsetooth Reservoir spelled doom for the town of Stout. To get close to one of the historic Stout quarries, take a short walk south from the terminus of the Shoreline Trail in Lory SP.

The invention and application of cement in the early decades of the 1900s largely supplanted the use of the Lyons Sandstone for construction purposes. In fact today, just east of Lyons town center on SH 66 is a Cemex plant—the largest industrial operation in Lyons. The quarrying of the Lyons sandstone somewhat revived and to this day continues to supply decorative rock for patios, sidewalks, walls and building facades. Noteworthy localities that have used the Lyons as construction material include: University of Colorado – Boulder; Visitors Center at Red Rocks Park; World Trade Center Memorial in NY; Memorial of 9/11 crash of UA flight 93 in Pennsylvania.

Environmental Impact

If you have traveled along Buckhorn Road (CR 27) between Masonville and US Route 34 you will have noticed the extensive quarrying of Lyons Sandstone along the west side of the road including at the entrance to Bobcat Ridge Natural Area. These historic quarries are operated by several different families and have changed ownership over the years. The aesthetics are not pleasing and the destruction of habitat unsettling, which begs the question, for what purpose?

Throughout history, dating back to at least the Egyptians and the pyramids, quarries have been a vital source of rock upon which our civilization is built. Rock remains an important resource for the construction of the buildings and roads we all depend on. However, much of the quarrying of the Lyons Sandstone is now dependent on more discretionary aesthetic choices we all make: fireplaces, pool accents, mailbox enclosures, paving stones, garden benches, and paving stones. We are confronted by these choices when we see firsthand the destruction that results.

Oil and Gas

There are only a handful of oil fields within the Denver Basin that produce from the Lyons Sandstone. It is the oldest formation in the basin productive of oil and gas. Two of the biggest fields discovered in the 1950s using seismic methods are located about 15 miles NE of Fort Collins in Weld, County. Named Black Hollow and Pierce Fields, both are structural anticlines that have produced 10.8 and 11.5 million barrels of oil respectively.

Geologic Interpretation

Geology, like all science, is subject to continuous updating based on new evidence or better interpretations that explain all of the facts. With time, more studies of analogs from around the world, new experimental data, and study of modern geologic processes build a general knowledge base from which to make better interpretations of the rock record. For example, the Lyons Sandstone was once thought to be an ancient marine beach deposit. But after intensive study of the sandstone in Sterling Quarry in Lyons, Colorado, geologists reinterpreted the Lyons Sandstone as an ancient desert dune deposit. Among the observations in favor of a wind-blown deposit are: well-rounded and sorted fine grained quartz grains; large scale (up to tens of feet thick) steeply dipping cross bedding; and fossil animal tracks and fossil raindrop impressions.

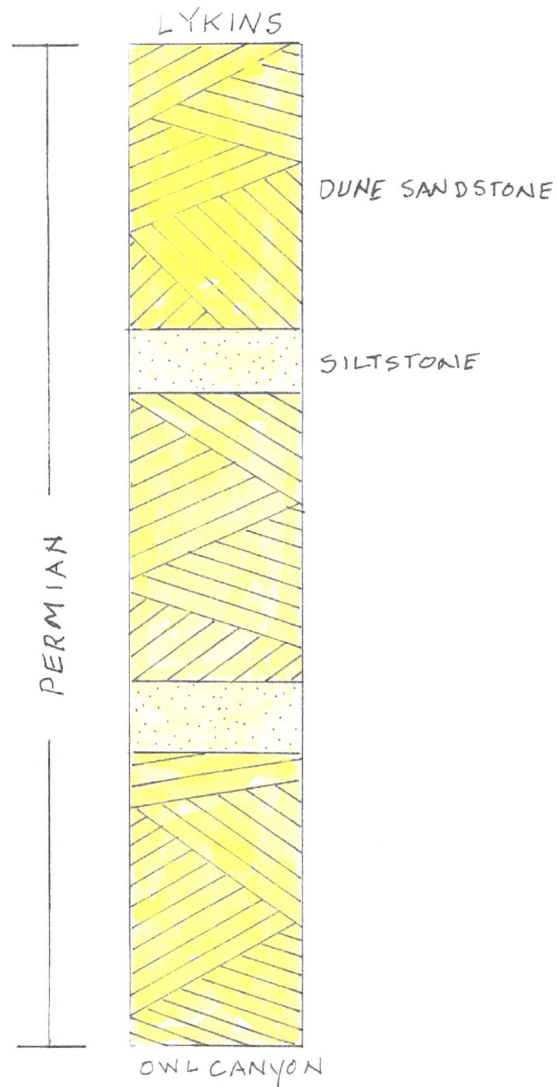

SCHEMATIC ROCK COLUMN
LYONS SANDSTONE

LYKINS

DUNE SANDSTONE

SILTSTONE

PERMIAN

OWL CANYON

Fig. 33 The large scale cross bedding of the Lyons Sandstone reflects deposition as ancient sand dunes. Thin, horizontally bedded siltstones represent interdune deposits. The Lyons gradually thins to a feather edge near the Colorado-Wyoming border.

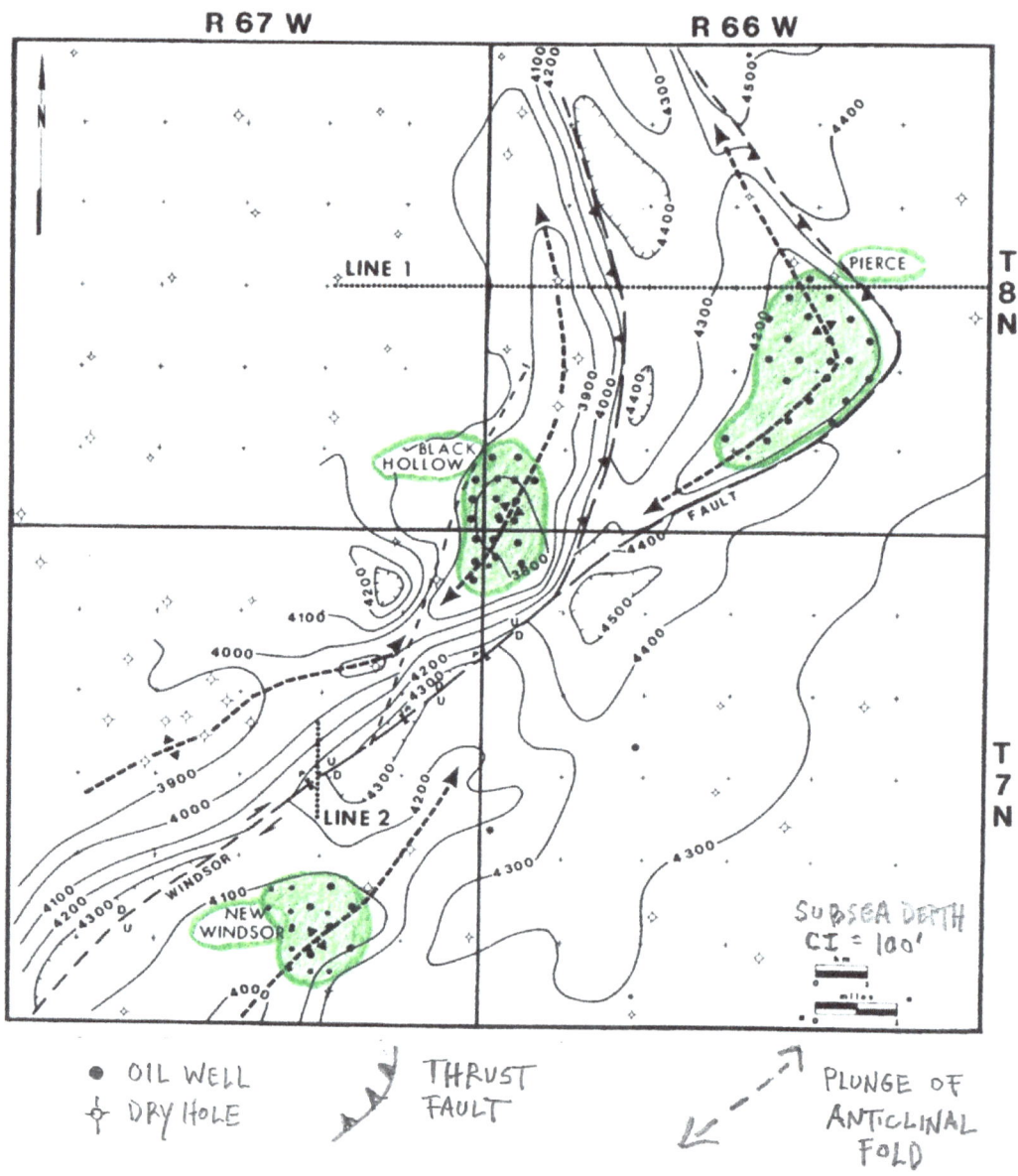

Fig. 34 Lyons subsurface structure map covering three Lyons Sandstone oilfields in Weld County, Colorado. The oilfields are located over 9,000 feet below the surface and were discovered using seismic technology in the 1950s. Further west in Larimer County, the Wellington and Fort Collins oilfields also produce from subsurface anticlines, but the reservoir is the Muddy Sandstone Member of the Dakota at a depth of about 4,000 feet. They were discovered in the 1920s using surface mapping techniques. (Modified from Stone DS, 1985)

Fig. 35a Historic Lyons Sandstone Quarry at Horsetooth Reservoir in Lory State Park.

Fig. 35b Active Lyons Sandstone Quarry at entrance to Bobcat Ridge NA.

Fig. 35c Cross bedding in Lyons Sandstone at an outcrop just off the Shoreline Trail in Lory State Park.

Fig. 35d Contact of tan colored Lyons Sandstone on the ridge with red- bedded Owl Canyon rocks forming the slope below. Entrance road into Bobcat Ridge Natural Area.

Fig. 35e Outcrop of cross-bedded Lyons Sandstone off of CR 25E, southwest of Bellvue.

Fig. 35f Profile view of the Lyons hogback through Mill Canyon on CR 25E southwest of Bellvue.

LYKINS FORMATION

AGE: 248—275 million years ago (Late Permian—Early Triassic)

THICKNESS: 600—800 feet

DEPOSITIONAL ENVIRONMENT: Arid coastal alluvial plain, tidal mud flat, shallow marine

LITHOLOGY: Mudstone, limestone, gypsum

WHERE ARE GOOD EXPOSURES OF THE LYKINS?

Red Mountain Open Space:
    Bent Rock Trail - Giant blocks of quarried white **gypsum** and **alabaster** can be seen just off the south segment of the trail.
    K-Lynn Cameron Trail - Isolated exposures of **stromatolite** fossils within the Forelle Limestone can be found on the lower reaches of the trail.
    Sinking Sun Trail - Lykins outcrops all along trail.

Lory State Park:
    Shoreline Trail to Horsetooth Reservoir - When the reservoir is drawn down there are several small headlands that emerge along the west shoreline that expose eroded Forelle Limestone outcrops littered with stromatolite fragments, and **leached** and **brecciated** fragments of gypsum.

Devil's Backbone Open Space:
    Wild LoopTrail - Exposure of the Lykins underlays the east segment of the Loop.

HOW DO I IDENTIFY THE LYKINS?

- Forms the valley between the Lyons (west) and Dakota (east) hogbacks
- Predominantly brick red crumbly mudstone with thin beds of consolidated siltstone
- Gypsum deposits near the base—often leached and brecciated
- Thin interlayers of limestone often with laminated (crinkly) stromatolite structure

---

TOPICS OF INTEREST:

Valley Landscape

The dominant mudstone lithology of the Lykins erodes easily and forms a long north-south oriented **strike valley** within the Foothills outcrop belt. The Lykins valley attracted early homesteaders as a desirable location for farming and ranching. The low topography bounded by the Lyons and Dakota hogbacks also serves as a natural container for impounding giant water reservoirs such as at Horsetooth, Carter Lake and the proposed Glade Reservoir north of Ted's Place. A significant geologic risk became apparent when gypsum beds at the base of the Lykins began to dissolve, creating a leak under Horsetooth Dam. Horsetooth Reservoir was drained to about 5% capacity while a repair project was completed between 2001—03.

Gypsum Mining and Use

Gypsum was discovered within the Lykins formation during the construction of the Louden irrigation ditch near Devils Backbone in 1888. A Loveland business called the Consolidated Plaster Company was soon formed and the gypsum beds were ultimately mined to a depth of over 100 feet. Although the mine is no longer active, the tailings and mine pit are clearly visible from the Wild Loop Trail at Devil's Backbone Open Space.

The mineral gypsum is a hydrated calcium sulfate ($CaSO_4\ 2H_2O$). Its primary industrial and economic value is in the manufacture of drywall. Gypsum is formed through precipitation from shallow hypersaline waters in arid environments.

A fine grained form of gypsum known as alabaster was later discovered in Owl Canyon. In the 1930s and 40s several companies and quarries supported a burgeoning Colorado alabaster sculpture industry that attained world renown for production of lighthouse lamps, statues, vases, and bowls. Although the early quarries have gone out of business, the Colorado Alabaster Company located in Fort Collins operates the Munroe Quarry northwest of the Park Creek Reservoir outside of Livermore. The gypsum is mined for use as amendments for cement and agricultural soil, while the alabaster is sold for artistic use.

Stromatolites

Stromatolites are sedimentary structures produced by one of the oldest known lifeforms on earth, **cyanobacteria**. When cyanobacteria, which form a sticky biofilm, colonize a sedimentary surface they trap sediment, and precipitate limestone from marine water as a product of photosynthesis. The structures build up as the cyanobacteria continually grow up through successive layers of new sediment. A wide variety of stromatolite shapes and sizes result including broad domes tens of meters wide down to centimeter scale structures. In cross-sectional view the structures appear in limestones as wavy, thin **laminations**.

Stromatolites have been identified in rocks dated as old as 3.5 billion years. Photosynthetic cyanobacteria became so prevalent that they led to the development of free oxygen in the earth's atmosphere. A key pivot point in earth's history, the named Great Oxidation Event is thought to have occurred around 2.4 billion years ago.

Stromatolites, such as those found in the Forelle limestone most likely formed in a hypersaline sea where grazing predators were not able to survive. Stromatolites are still forming today in hypersaline environments like Sharks Bay, Australia.

Permian Extinction — 'The Great Dying'

The greatest extinction event known in the geologic record, where approximately 90% of all marine and 70% of all terrestrial species went extinct is thought to have happened at about the time the Poudre Limestone member of the Lykins Formation was being deposited 252 million years ago. At the time, all of earth's continents were assembled into the giant supercontinent Pangea. In the northern part of Pangea, a massive outpouring of basalt known as the Siberian Flood Basalts is commonly accepted as the trigger for the great die-off. It was not the basalt itself, but the associated expelled gases that had life killing consequences. Carbon dioxide, a well-known green-house gas likely caused sea-surface temperatures to rise by as much as 18 degrees Fahrenheit. The increase in sea temperatures resulted in depletion of dissolved oxygen critical for support of life. Other gases such as chlorine and fluorine are thought to have destroyed 70% of the ozone shield. The Permian extinction was another pivotal event for evolution of life on earth.

SCHEMATIC ROCK COLUMN

LYKINS FORMATION

JELM

MEMBER NAMES

RED HILL SHALE

TRIASSIC

PARK CREEK LIMESTONE
STONEWALL CREEK SHALE
POUDRE LIMESTONE
LIVERMORE SHALE

FORELLE LIMESTONE

GLENDO SHALE (GYPSUM at BASE)

FALCON TONGUE LIMESTONE

HARRIMAN SHALE
    (GYPSUM at BASE)

PERMIAN

LYONS

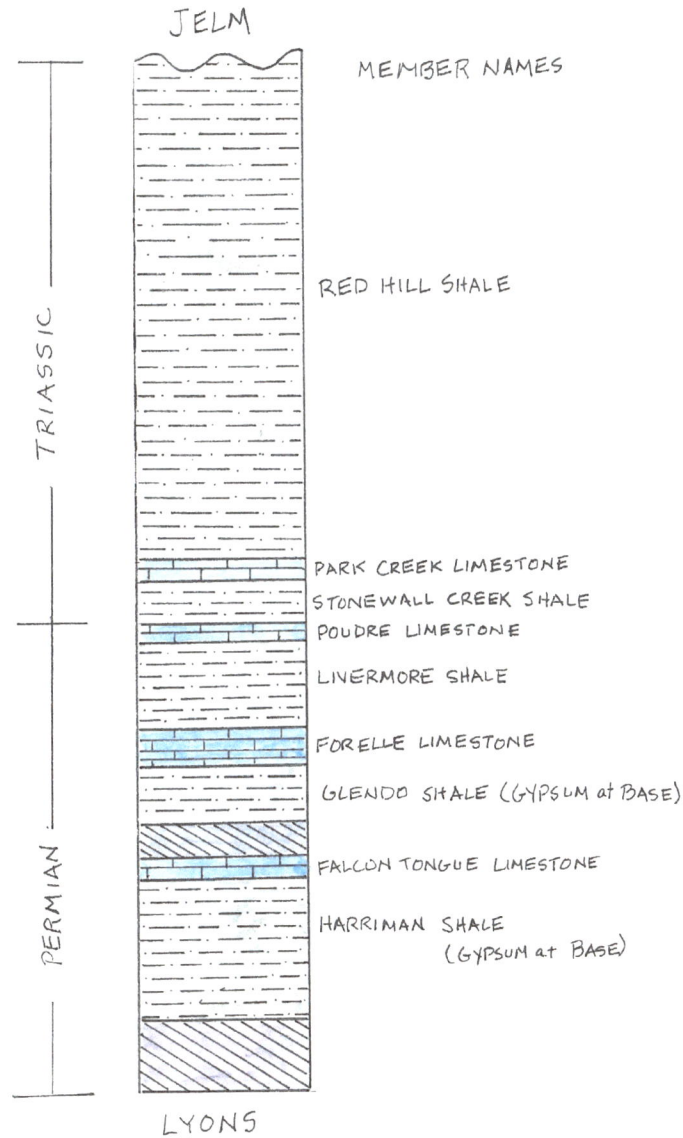

Fig. 36 Dominantly a mudstone and shale deposited on a broad tidal-flat. The lower Permian section of the Lykins however is punctuated with marine limestones bearing stromatolites as well as gypsum beds that were precipitated as the hypersaline sea evaporated.

DEVELOPMENT OF STROMATOLITES

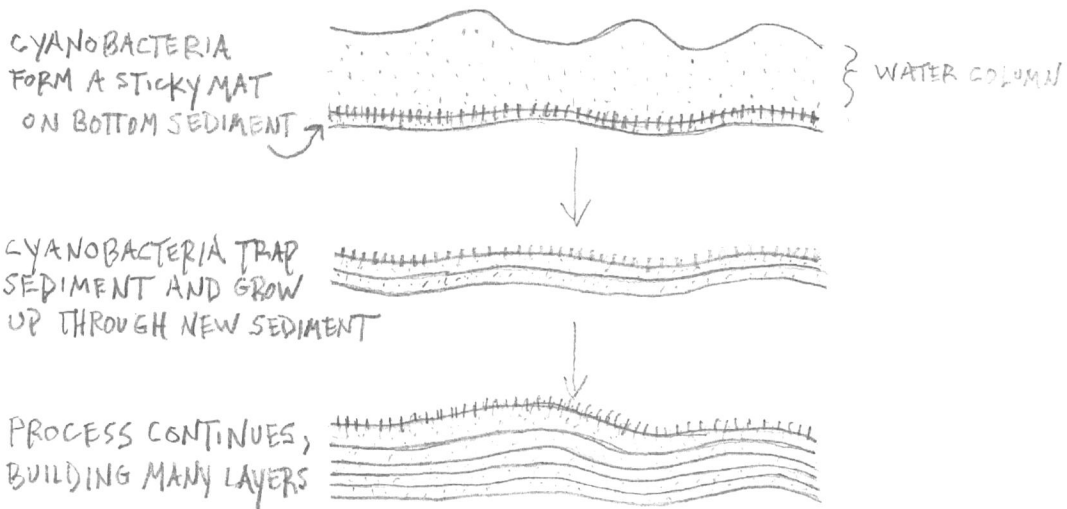

CYANOBACTERIA
FORM A STICKY MAT
ON BOTTOM SEDIMENT

WATER COLUMN

CYANOBACTERIA TRAP
SEDIMENT AND GROW
UP THROUGH NEW SEDIMENT

PROCESS CONTINUES,
BUILDING MANY LAYERS

PLAN AND PROFILE VIEWS OF PRESERVED
STROMATOLITES IN A ROCK OUTCROP

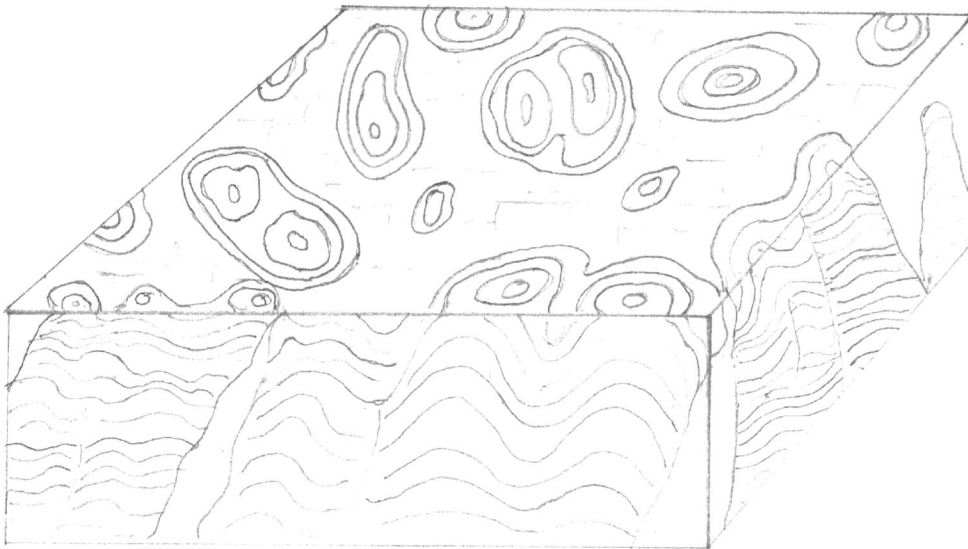

MBK 2021

Fig. 37a & b Schematic of how stromatolites form and how they appear in the rock record.
(a. Modified from Earth System History, Freeman, 2005;
b. stromatolitesgl2019.wordpress.com)

Fig. 38a Cyanobacteria, perhaps the oldest and most resilient lifeform on earth. Stromatolites in the rock record indicate they have been around for over 3 billion years. (stromatolitesgl2019.wordpress.com)

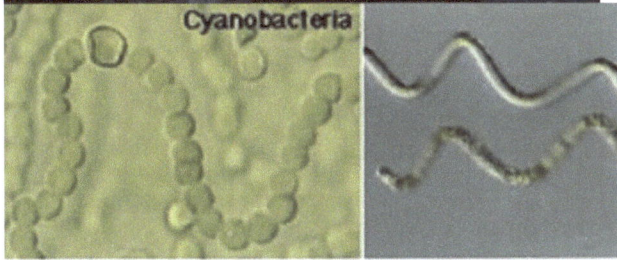

Fig. 38b Modern stromatolites from Sharks Bay, Australia. (oceanpark.com.au)

Fig. 38c Top view of a stromatolite outcrop within the Forelle Limestone Member of the Lykins Formation at Red Mountain Open Space.

Fig. 39a When Horsetooth Reservoir is drawn down, exposed headlands on the west side are littered with blocks of gypsum breccia and stromatolitic limestone.

Fig. 39b Blocks of alabaster—a fine grained form of gypsum—are seen near an abandoned quarry just off the south segment of Bent Rock Trail at Red Mountain Open Space.

Fig. 39c Gypsum and alabaster beds exposed on Big Hole Wash Trail at Red Mountain OS.

Fig. 39d The Monroe Quarry, an active gypsum mining operation located just south of Red Mountain Open Space.

Fig. 39e A thin white layer of Lyons Sandstone separates the similarly colored Lykins Fm above from the Owl Canyon Fm below. Table Mountain in background. Bent Rock Trail at Red Mountain OS.

Fig. 39f Red beds of the Lykins Fm are seen here dipping about 25 degrees to the west. They are part of the west limb of the Sand Creek Anticline, the crest of which has been eroded off. North view from the Big Hole Wash Trail.

Fig. 39g Lykins redbeds exposed along West CR 38E on the southeast side of Horsetooth Reservoir.

JELM & SUNDANCE FORMATIONS

AGE: Jelm 240—248 million years ago (Upper Triassic)
     Sundance 165—170 million years ago (Mid-Upper Jurassic)

THICKNESS: 150—200 feet

DEPOSITIONAL ENVIRONMENT: Eolian sand dunes and shallow marine

LITHOLOGY: Sandstone

WHERE ARE GOOD EXPOSURES OF THE SUNDANCE AND JELM?

The USGS maps these two formations undivided in the northern Front Range foothills.

Devil's Backbone Open Space:
  Morrison Trail - A worthwhile geologic interpretive trail, the Sundance Fm is exposed right below your feet until the trail turns west over the Morrison Fm.
  Wild Loop Trail - Exposure of the Jelm and Sundance along the south segment of the loop.

Red Mountain Open Space:
  Big Hole Wash Trail - Jelm and Sundance are exposed on the slope-face of the topographic high west of the trail.
  CR 21 - Perhaps the best exposures of the Jelm and Sundance Formations can be seen on the east side of CR 21 on the way into Red Mountain OS. Look for the colors noted below on the lower to mid slopes beneath the Dakota hogback.

HOW DO I IDENTIFY THE SUNDANCE/JELM?

- Forms the lower slopes on the west face of the Dakota hogback
- Sundance is yellow to gray in color at Red Mountain, but red/brown at Devil's Backbone
- Jelm is pale red color, overlies deeply red colored Lykins Formation
- Cross-bedding evident on Morrison and Wild Loop Trails

TOPICS OF INTEREST:

Sand Seas

Geologists refer to rocks formed from deposition of sand by wind as **eolian** deposits—named after Aeolus, the Greek god of winds. In arid environments with a good supply of sand and steady wind, the deposits can develop into migrating dunes. Two primary environments of deposition are 1) coastal dunes that many of us are familiar with near the beach, and 2) vast inland dune fields (ie., sand seas, or Ergs) such as in the Sahara desert.

The Colorado Plateau holds the largest concentration of eolian sedimentary rocks in the geological record. Sometimes referred to as 'the great sandpile' these ancient sand dunes reflect the past existence of extensive sand seas and coastal dune environments. Development of these vast desert dune fields occurred during the Pennsylvanian through Jurassic geologic periods, a time when this part of the western interior was located in subtropical latitudes known as desert climate zones.

In the northern Front Range foothills, there are four rock units that partly include or are mostly composed of eolian sandstones. They were deposited on the eastern margin of the vast desert dune region of the Colorado Plateau. They are the:

Ingleside Formation - Cross-bedded dune deposits at the top of the formation
Lyons Sandstone - Predominantly an inland dune deposit
Jelm and Sundance Formations - Both thought to be preserved remnants of inland sand seas

Major Jurassic Unconformity

Although the Jelm and Sundance Formations are both interpreted as ancient dune deposits in the northern Front Range foothills, this similarity belies the fact that there are missing pages from the geologic story between the deposition of these two formations. In exposures at RMOS there is a noted color change from the pale red Jelm Fm. to the yellow-gray Sundance Fm. But what is not readily apparent is that there is a major unconformity, or gap in the rock record between these two formations. Because the bedding is parallel between the two formations, there is no easy visual clue. But geologists who have regionally mapped and dated these rock formations have discovered there is in fact a 70 million year gap in the rock record between the Jelm and Sundance Formations. Only the Great Unconformity below the Fountain Formation represents a bigger time gap in the rock record as recorded in the northern Front Range foothills. The rocks that are missing due to erosion or non-deposition from the Front Range foothills rock record are abundantly present on the Colorado Plateau and range from the Chinle through the Carmel Formations. This includes the massive Navajo Sandstone, thought to record the most extensive desert dune system in the history of earth.

See the accompanying figure for a correlative comparison of the rock record from the Colorado Plateau with the Front Range foothills during the Permian, Triassic, and Jurassic Periods.

Pangaea Breaks Up

The supercontinent of Pangaea existed from about 320 to 180 million years ago, a time during which it is thought that 85% of earth's landmasses were united into one giant continent. All the sedimentary rocks in the northern Front Range foothills from the Fountain Formation through the Jelm Formation were deposited during this time reflecting the influence of the Ancestral Rocky Mountains (formed by the assembly of Pangaea), broad floodplains, marginal seas, and periods of arid desert conditions.

During the time of the Jurassic Unconformity discussed above, Pangaea began to split apart about 180 million years ago. Rifting and sea floor spreading separated North America from Africa creating the Atlantic Ocean which continues to expand to this day. Simultaneously the western margin of North America became an active subduction zone leading to compressional stresses resulting in **fold and thrust** mountain belts extending into central Utah and flooding of the continental interior by seaways. The ancient sand dunes of the Sundance Formation were deposited along the southern margin of one of the first marine incursions onto the continent known as the Sundance Seaway.

Thus, the Jurassic Unconformity serves as a major transitional marker from rocks that were deposited as part of Pangaea to rocks that were deposited within or adjacent to interior seaways, and fold and thrust mountains. The Jelm and Sundance Formations lay across this divide.

SCHEMATIC ROCK COLUMN
JELM AND SUNDANCE
FORMATIONS

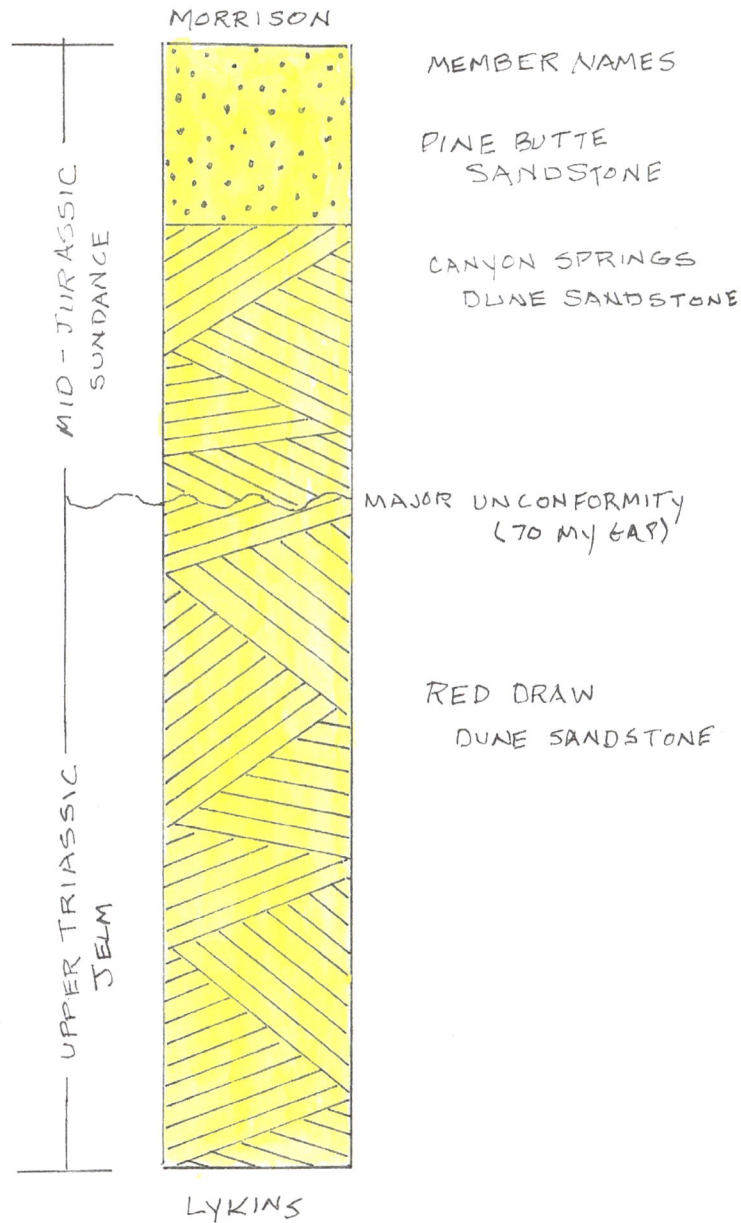

MORRISON

MEMBER NAMES

PINE BUTTE
SANDSTONE

CANYON SPRINGS
DUNE SANDSTONE

MID - JURASSIC
SUNDANCE

MAJOR UNCONFORMITY
(70 MY GAP)

RED DRAW
DUNE SANDSTONE

UPPER TRIASSIC
JELM

LYKINS

Fig. 40 The eolian sandstones of the Jelm and Sundance formations are sandwiched between the mudstones of the Morrison and Lykins formations. Although deposited in similar desert dune environments there is in fact a 70 my gap in time between the Jelm and Sundance formations.

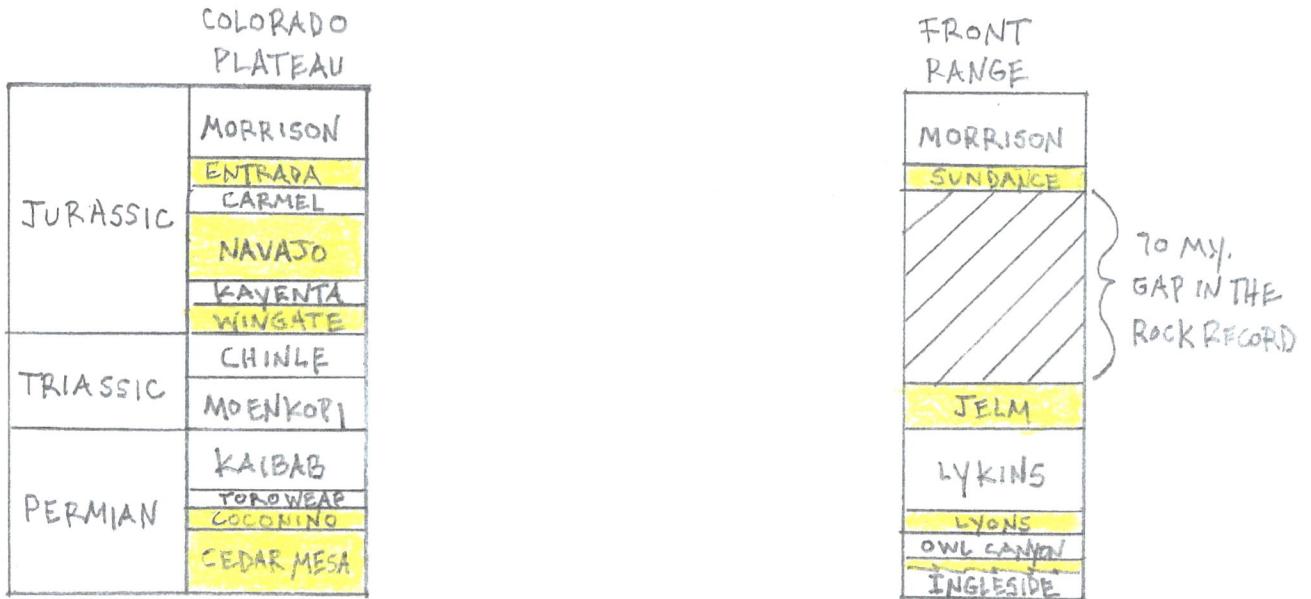

Colorado Plateau

JURASSIC	MORRISON
	ENTRADA
	CARMEL
	NAVAJO
	KAYENTA
	WINGATE
TRIASSIC	CHINLE
	MOENKOPI
PERMIAN	KAIBAB
	TOROWEAP
	COCONINO
	CEDAR MESA

Front Range

MORRISON
SUNDANCE
70 MY. GAP IN THE ROCK RECORD
JELM
LYKINS
LYONS
OWL CANYON
INGLESIDE

Fig. 41 Rock record comparison between the Colorado Plateau and the Northern Front Range foothills during the period of widespread desert dune deposits. Rocks of the Upper Triassic through most of the Jurassic are not present in the Front Range foothills.

Fig. 42 At Red Mountain OS the orange colored Jelm and yellow/gray colored Sundance formations are often well exposed on the slope beneath the Dakota hogback ridge.

Fig. 43 The Jelm and Sundance Formations are sandstones that were deposited as an inland dune field or sand sea during the Triassic and Jurassic periods. They may have looked much like this image from Great Sand Dunes National Park. (nps.gov)

Fig. 44a Cross-bedded Sundance Fm on the Morrison Trail at Devil's Backbone.

Fig. 44b One of the best and most accessible exposures of the Sundance Fm is alongside West CR 38E south of Spring Canyon Dam. Here, the tabular yellow colored beds are marine sandstones of the Pine Butte Member that overlay the cross-bedded eolian sandstone of the Canyon Springs Member.

Fig.44c Subtle cross-bedding in the Jelm sandstone on the Wild Loop Trail at Devil's Backbone.

Fig. 44d Better exposures of the Jelm Fm can be found on the east flank of the Milner Mountain Anticline adjacent to the Louden Ditch. Here the cross-bedding is more apparent. (The site is not open to the public at this time)

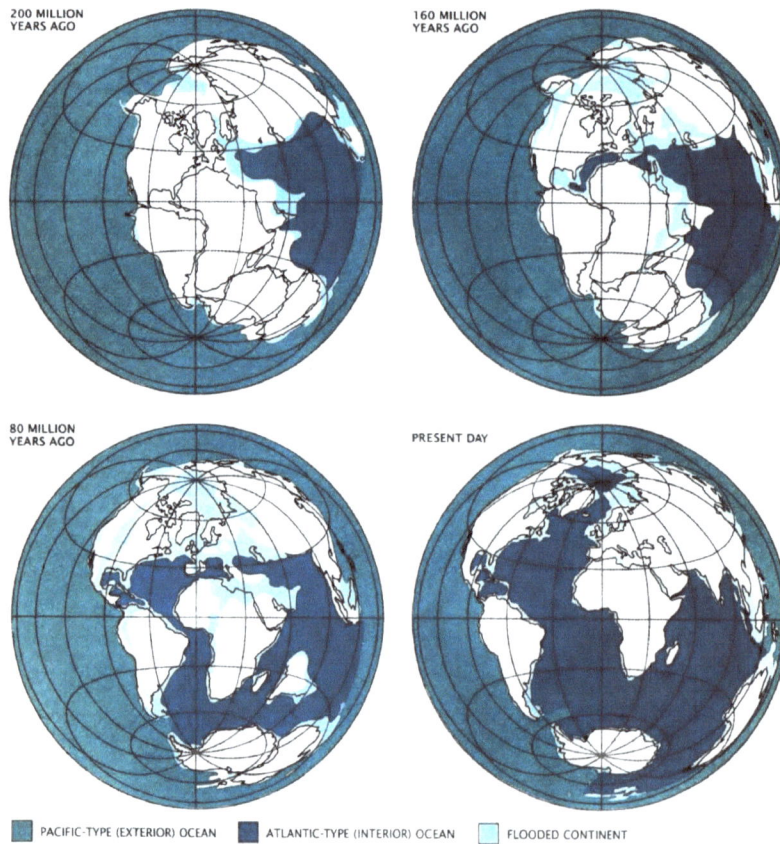

200 MILLION YEARS AGO

160 MILLION YEARS AGO

80 MILLION YEARS AGO

PRESENT DAY

PACIFIC-TYPE (EXTERIOR) OCEAN   ATLANTIC-TYPE (INTERIOR) OCEAN   FLOODED CONTINENT

Fig. 45 The supercontinent Pangaea began assembly about 300 mya, and then began breaking apart about 170 mya. The illustration above captures four snapshots in time showing the positions of the continents. Deposition of the Fountain through the Jelm formations occurred during the period of Pangaea's existence as represented in the 200 mya view. Deposition of the Sundance and Morrison formations occurred during the early stages of breakup as represented in the 160 mya view. And deposition of the Dakota through Pierre Shale occurred during the period of marine flooding of western North America as represented in the 80 mya view. (Larson, Scientific American, 1995)

MORRISON FORMATION

AGE: 148—157 million years ago (Upper Jurassic)

THICKNESS: 320 feet

DEPOSITIONAL ENVIRONMENT: River and floodplain, shallow wetland lakes

LITHOLOGY: Sandstone, mudstone, claystone, and limestone

WHERE ARE GOOD EXPOSURES OF THE MORRISON?

Devil's Backbone:
Wild Loop Trail - At the Overlook location, gray limestone beds of the Morrison Fm protrude from the surface; and at the Keyhole, the contact with the Dakota can be observed. The contact surface represents a 46 million year gap in the rock record between the two formations. At the contact, the Morrison is a variegated mudstone.

Morrison Trail - About halfway up the trail you'll note the change from the reddish brown rocks of the Sundance Fm to the gray rocks of the Morrison Fm.

US 287 Roadcut (~3/4 mile north of Ted's Place):
- Exposure of the Morrison on the south side of US 287 as it curves eastward.

HOW DO I IDENTIFY THE MORRISON?

- Forms the west-facing upper slope directly below the Dakota hogback ridge
- Gray limestone beds and mudstone associated with freshwater lake or wetlands deposits
- Floodplain mudstones often exhibit variegated colors of green and red indicative of ancient soil development (paleosols)

TOPICS OF INTEREST:

The Vast Morrison Basin & Dinosaurs

One of the most impressive features of the Morrison Formation is its geographical extent; covering parts or all of 13 western states, from northern New Mexico to southern Canada and from eastern Utah to western South Dakota. Such a wide area might usually be associated with marine deposits, but the Morrison was in fact deposited on a vast alluvial plain dotted with wetlands and lakes. Most of the river systems were east flowing, draining western uplands representing the beginnings of the **Sevier Orogeny**. In the Colorado Front Range area, the Morrison is represented by carbonate wetland and lake deposits interbedded with floodplain mudstones, paleosols, and channel fill deposits.

Dinosaurs were the dominant creatures on the Morrison alluvial plain, and occasionally they would be swept up in flash floods that concentrated their remains in bone bed deposits, ultimately to be discovered by humans 150 million years later.

In 1877, Arthur Lakes, an English emigrant who taught geology at the newly minted Colorado School of Mines, discovered the world's first Stegosaurus fossil within the Morrison Formation near Morrison, Colorado. With the financial support of Othniel Charles Marsh, America's first professor of paleontology at Yale, Lakes developed excavations at thirteen quarries that yielded several more 'first' discoveries including an Apatosaurus (formerly called a Brontosaurus). At Dinosaur Ridge in Morrison, Colorado one can still see actual dinosaur bones embedded in the Morrison Formation at the bone quarry exhibit.

More recently and locally, bones of a rare dinosaur were discovered in 1990 in Masonville, Colorado by a University of Colorado graduate student. Referred to as the 'Masonville Monster' the bone pieces suggest the dinosaur was 50 feet long and weighed four tons. Paleontologists classify it as Epanterias, a species of the allosaur family. Allosaurs, like the Tyrannosaurs who lived later during the Cretaceous Period were ferocious meat-eaters that walked on their hind legs. The Masonville Monster bones are now housed at the Denver Museum of Nature, but replicas are also kept at the Fort Collins Museum of Discovery.

Uranium/Vanadium Deposits

Many dinosaur bones have been found to be radioactive. Sometimes hot enough that museums have covered their displayed skeletons with leaded paint, and academic paleontologists or commercial dinosaur hunters use portable geiger counters to locate bone deposits. But, more broadly, sandstone deposits within the Morrison Fm have been commercially mined for uranium and vanadium minerals throughout the 20th century.

The primary source of uranium and vanadium is believed to be volcanic ash-fall deposits (known as tuffs) that are interbedded within the Morrison. Oxidized uranium and vanadium within the tuffs is highly soluble and likely worked its way through groundwater to highly permeable channel sand deposits within the Morrison. When reaching a zone depleted in oxygen such as those containing organic matter, the oxidized uranium and vanadium becomes reduced and precipitates out, coating sand grains, plant material, and even dinosaur bones. One of the precipitates is a soft powdery mineral known as carnotite, containing both uranium and vanadium. Carnotite's eye catching yellow color may have lured Ute and Navajo Indians to apply the radioactive mineral as body paint.

Uranium and vanadium ores associated with the Morrison are concentrated on the Colorado Plateau in southwest Colorado where more than 1200 mines were once in operation. Uranium's primary use is for nuclear fuel while vanadium is used in specialty steel alloys for the aerospace, bicycle, defense, and tool industries.

Sub Dakota Unconformity

The top of the Morrison Formation is marked by a significant gap in the rock record with the overlying Dakota Group of rocks. Known as the Sub Dakota Unconformity, it represents a time interval of approximately 46 million years after which there is a significant environmental change to marine conditions with the flooding of the interior of the continent by the Western Interior Seaway.

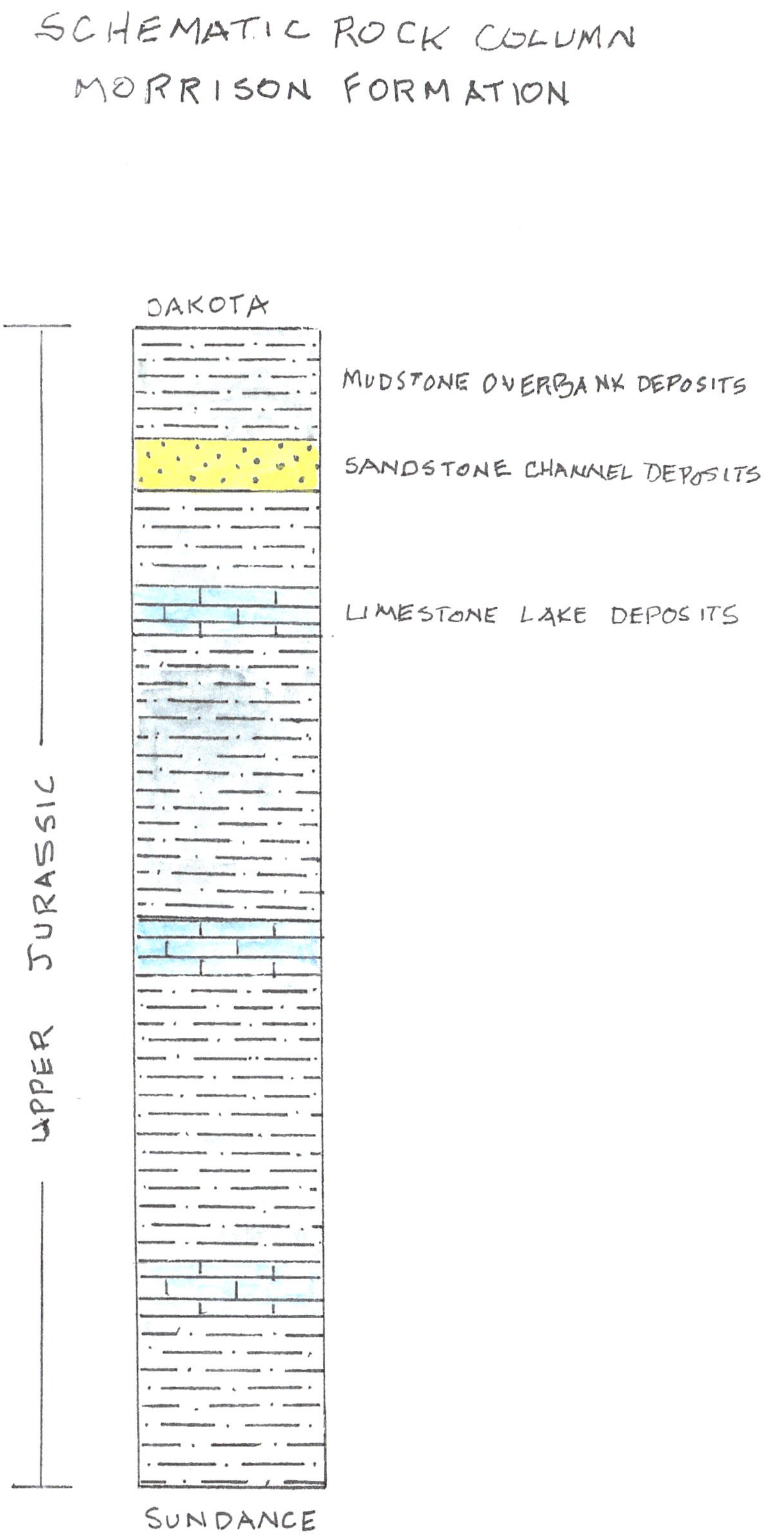

SCHEMATIC ROCK COLUMN
MORRISON FORMATION

DAKOTA

MUDSTONE OVERBANK DEPOSITS

SANDSTONE CHANNEL DEPOSITS

LIMESTONE LAKE DEPOSITS

UPPER JURASSIC

SUNDANCE

Fig. 46 The Morrison Formation is dominated by alluvial plain mudstone interbedded with sandy channels, and limestone deposited in wetland lakes.

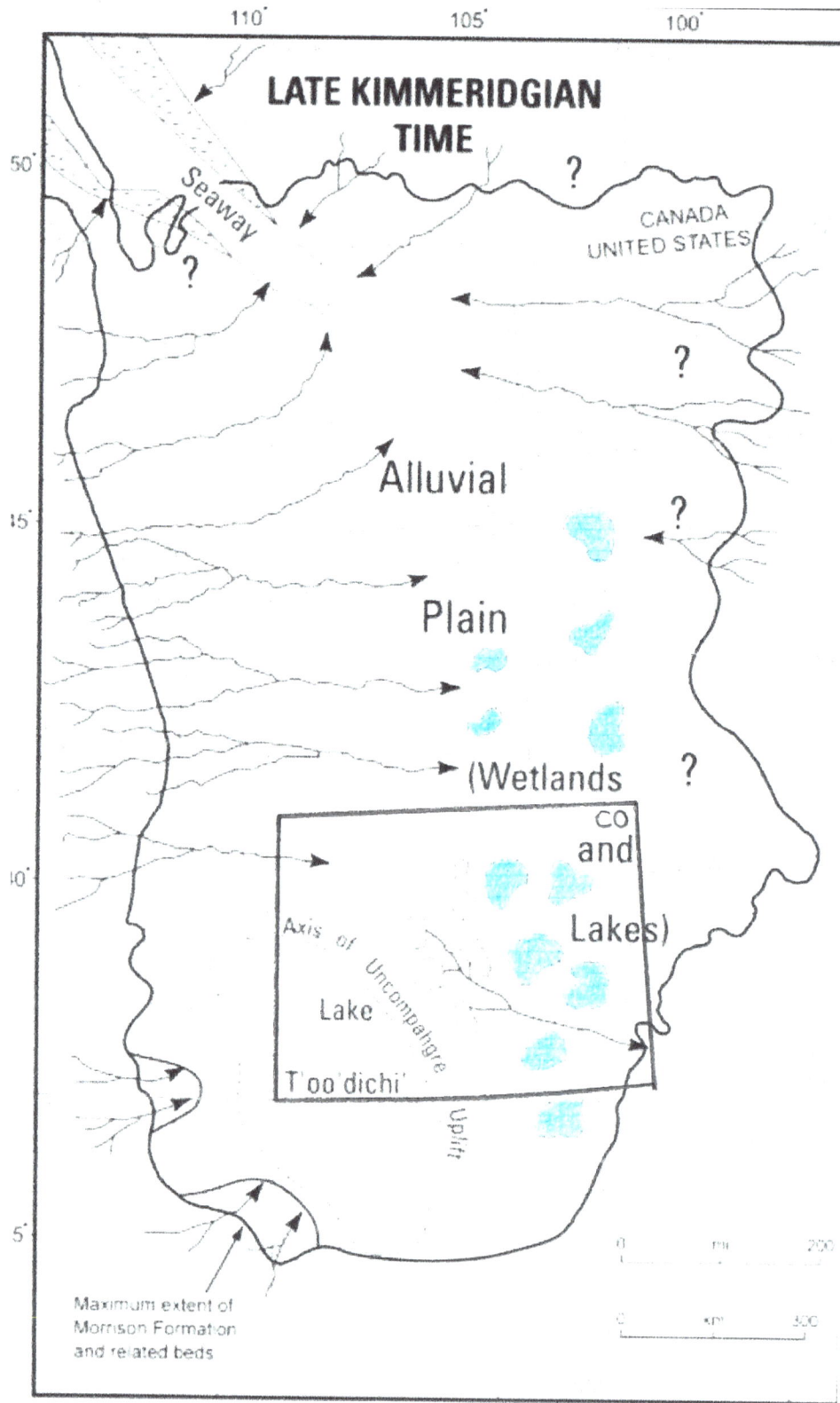

Fig. 47 Paleogeography of the extensive Morrison Formation.
(Turner, Sedimentary Geology, 2004)

Fig. 48a Dinosaur bones embedded in a sandstone channel deposit within the Morrison Formation. Dinosaur Ridge in Morrison, Colorado.

Fig. 48b Stegosaurus—first discovered in Morrison, Colorado in 1877— is Colorado's state fossil. (Wikipedia)

Mounted skeleton of Stegosaurus stenops at the Natural History Museum, London. Creative Commons License

Fig. 48c Dinosaur bone mineralized with radioactive carnotite.
(CSM Geology Museum)

Fig.49a Historical Morrison Formation quarry visible from Devil's Backbone parking lot. The Morrison occupies the narrow sloping valley between the Dakota "backbone" on the left and a low-relief ridge on the right formed by the Sundance and Jelm formations.

Fig. 49b Outcrop of steeply dipping Morrison Fm beds on the Morrison Trail at Devil's Backbone.

Fig. 49c Gray limestone and shale of the Morrison Fm outcrops along West CR 38E on the southeast side of Horsetooth Reservoir, south of Spring Canyon Dam.

Fig. 49d Roadcut exposure of the Morrison Fm on US 287 just north of Ted's Place. Mostly an interbedded limestone/shale sequence.

DAKOTA GROUP

AGE: 102—98 million years ago (Early Cretaceous)

THICKNESS: 300 feet

DEPOSITIONAL ENVIRONMENT: River, beach, delta, and offshore marine

LITHOLOGY: Sandstone, conglomerate, and shale

WHERE ARE GOOD EXPOSURES OF THE DAKOTA?

Coyote Ridge:
  Coyote Ridge Trail - The Dakota hogback is double-humped. The Muddy Sandstone forms the eastern ridge, followed by a valley underlain by the Skull Creek Shale before ascending to the top of the second ridge formed by the Plainview Sandstone.

Reservoir Ridge:
  Foothills Trail - Hummocky topography formed by Muddy Sandstone rockslide.
  North Loop Trail - Outcrops of Plainview Sandstone and Lyle Fm on west segment of loop.

Devil's Backbone:
  Wild Loop Trail - Steeply dipping beds of the Lytle Fm form the west limb of the Milner Mountain anticline (Fig. 57).

Maxwell:
  Foothills Trail - Muddy Sandstone rockslide features and hogback ridge sandstone.

Pine Ridge:
  Viewpoint Spur Trail - Muddy Sandstone outcrops with locally abundant trace fossils.
  Piano Boulders/Duncan's Ridge - Muddy Sandstone outcrops on the east, and Lytle Fm on the west side of Centennial Drive.

HOW DO I IDENTIFY THE DAKOTA?

- Buff colored sandstones forming hogback ridges
- Trace fossils and ripple marks common
- First prominent ridge rising from the Great Plains
- West-facing escarpment slopes littered with large toppled blocks from Dakota ridge top

---

TOPICS OF INTEREST:

Western Interior Seaway

The Dakota Group of rocks record an amazing transition of environments. The oldest rock unit in the Dakota Group—the Lytle Formation—was formed from coarse river borne deposits derived from the Sevier fold and thrust belt mountains in central Utah. The overlying Plainview Sandstone however is a marine deposit that records the initial incursion of the **Western Interior Seaway** that would persist for the next 30 million years. All the remaining younger rock units in the northern Colorado Front Range foothills were deposited within or marginal to this vast inland sea that finally receded with the rise of the Rocky Mountains 70 million years ago.

Dakota Hogback

Hogbacks are linear ridges made of weather resistant layers of tilted sedimentary rock. The Dakota Hogback is a renowned text-book example that extends for several hundred miles in a north-south direction from southern Wyoming through Colorado and into northern New Mexico. The conspicuous ridge is the first prominent foothill that neatly delineates the flat prairie of the Great Plains from the Rocky Mountains to the west.

In the Fort Collins area, the Dakota typically **dips** between 15 and 25 degrees down to the east. This dip slope is essentially a **bedding plane** (paleo depositional surface) and a good place to look for fossil ripple marks and **trace fossils**. The world famous dinosaur footprints discovered at Dinosaur Ridge in Morrison, Colorado are from the same interval of rocks. The steeper west side of the hogback is referred to as an escarpment, formed where the erosion surface cuts across the bedding into softer easily eroded shale. Centennial Drive on the east side of Horsetooth Reservoir lays on top of the Skull Creek Shale between the Muddy Sandstone hogback on the east and the Plainview Sandstone hogback on the west side.

Oil and Gas

The first oilfield west of the Mississippi River was accidentally discovered in 1862 while drilling a shallow water well near Florence, Colorado about three years after the famous Drake oil well discovery in Titusville, Pennsylvania. The oil was from shallow fractured Pierre Shale. The Florence-Canyon City Field as it is known today, still produces and is noted as the longest active oilfield in the country. The second oilfield in Colorado, Boulder Field, was discovered in 1901 (same year as Spindletop in Texas) in an area with known oil seeps and drilled in a location determined by **dowsing**. Thus, neither of the first two oilfields in Colorado were discovered with scientifically intentional methods.

Around 1915 however, based on surface mapping by professional geologists, it was determined that a subsurface anticlinal fold existed a few miles northeast of Fort Collins. By this time anticlines were established as potential natural hydrocarbon traps. In 1923 a successful well was drilled by the Union Oil Company of California producing from the Muddy Sandstone. Subsequent development wells established the Wellington and Fort Collins oil fields. These discoveries were significant in establishing production from a prolific new reservoir and pushing commercial production deeper and further east into the Denver Basin.

Rockslides—A Current Hazard?

An interesting aspect of the Dakota Hogback, especially along the stretch from Fort Collins north to Soapstone Prairie, is the common presence of **rockslides** that have occurred on the east dipping flank of the hogback. These are sometimes large features that can extend for up to a mile along the ridge with down dip movement of over 1000 feet. The slides are not of loose sediment, but rather large slabs of the upper Dakota Muddy Sandstone that became detached along a weak slip surface in the underlying Skull Creek shale. The slabs became disaggregated during down dip movement, forming an apron of hummocky topography. At Reservoir Ridge and Maxwell, you can walk over this hummocky surface which extends down slope to cover part of the valley underlain by Benton Group rocks. There are unknowns about whether the slides occurred catastrophically or by slow creep, and also about the current state of stress, and potential for new slides. The rockslides have been dated from 30 mya to several hundred thousand years ago—perhaps as little as even a few hundred years ago.

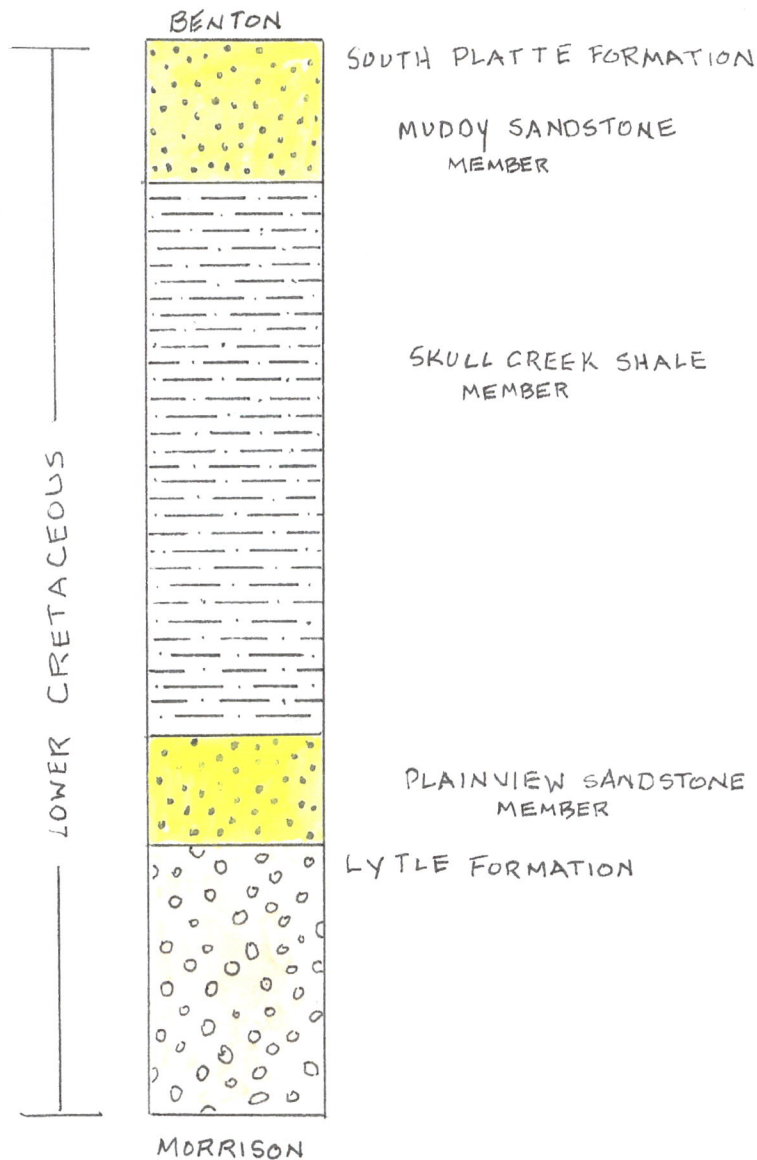

SCHEMATIC ROCK COLUMN
DAKOTA GROUP

BENTON

SOUTH PLATTE FORMATION

MUDDY SANDSTONE MEMBER

SKULL CREEK SHALE MEMBER

LOWER CRETACEOUS

PLAINVIEW SANDSTONE MEMBER

LYTLE FORMATION

MORRISON

Fig. 50 The Dakota Group comprises the South Platte Formation (with three members), and the Lytle Formation. Rocks of the Plainview Sandstone Member mark the onset of the Western Interior Seaway. The Lytle Formation is coarse-grained sandstone and conglomerate deposited by rivers.

Fig. 51
The Lytle Formation—oldest rock unit In the Dakota Group—was deposited by rivers as coarse conglomerates and sands derived from the Sevier fold-thrust belt in Utah. (Pang, 1995)

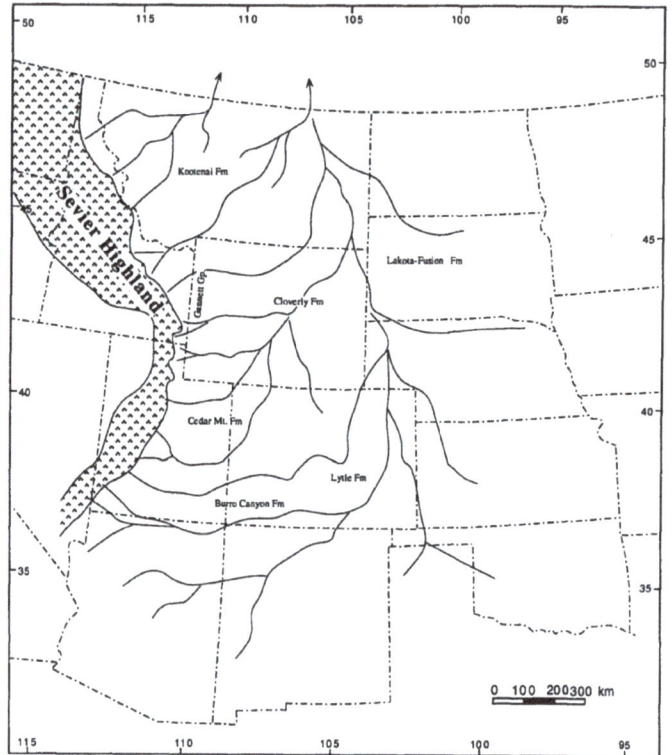

Fig. 52
After the Lytle Formation was deposited, a drastic change occurred with the incursion of the Western Interior Seaway. The seaway expanded and shrank over time but nonetheless persisted for the next 30 million years. All the rock units in the northern Colorado foothills starting with the Plainview Sandstone member of the Dakota up through the Pierre Shale were deposited within or marginal to this epeiric sea.
(cretaceous atlas.org)

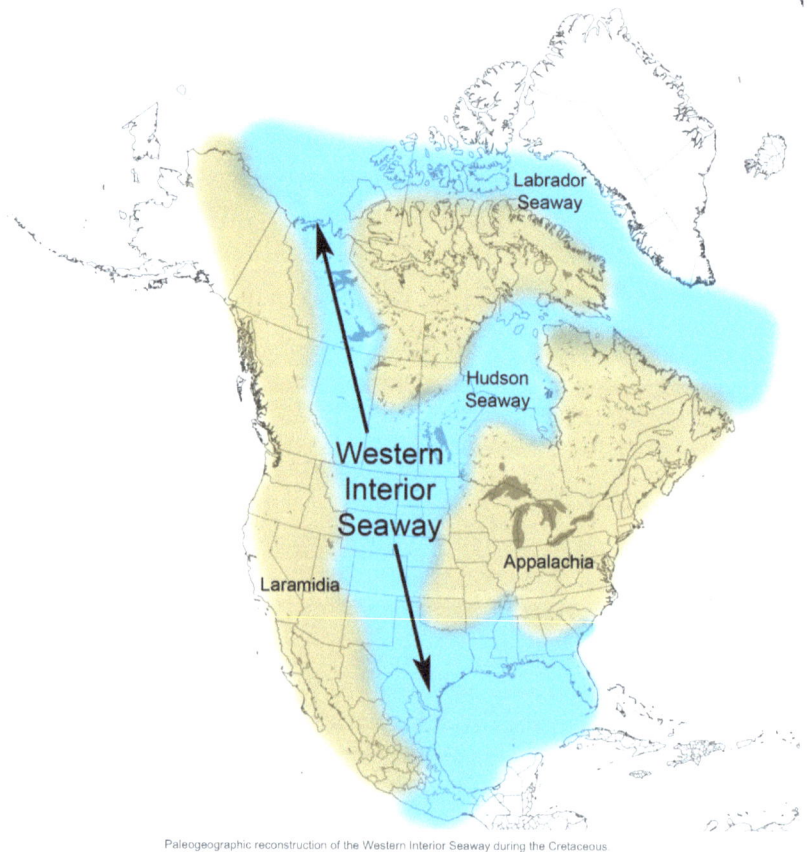

Paleogeographic reconstruction of the Western Interior Seaway during the Cretaceous

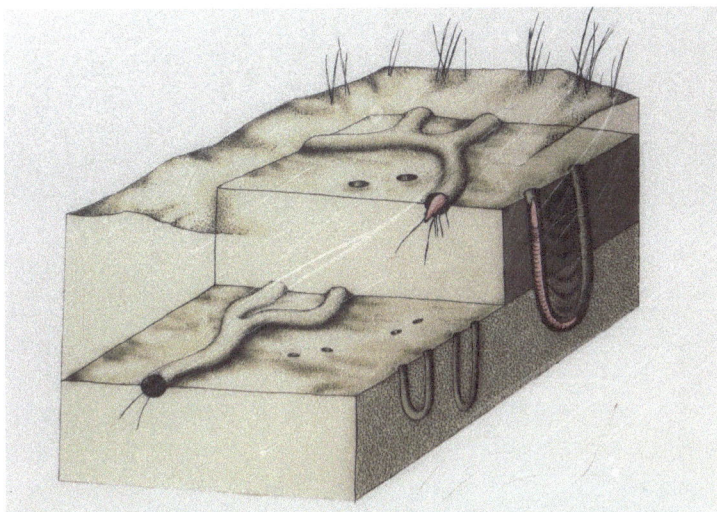
Fig. 53 Trace fossils formed by the burrowing activity of shrimp and lugworms are common n the Muddy Sandstone.
(Dinosaur Ridge interpretive sign)

Fig. 54a Trace fossil burrows seen on the top of a Muddy Sandstone boulder along the Viewpoint Spur Trail at Pineridge Natural Area.

Fig. 54b Dinosaur footprint in the Muddy Sandstone at Dinosaur Ridge in Morrison, Colorado.

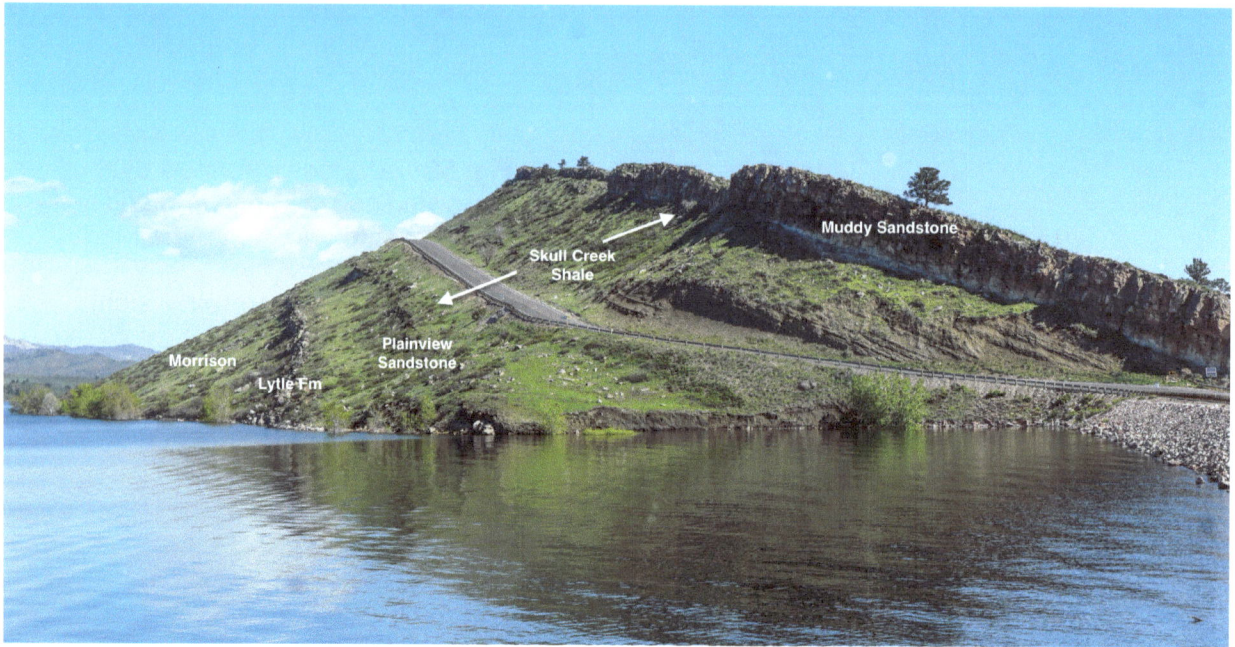

Fig. 55 View of the Dakota hogback on the north side of Dixon Dam at Horsetooth Reservoir. Rock units of the Dakota Group include—from youngest to oldest—the Muddy Sandstone down through the Lytle Formation.

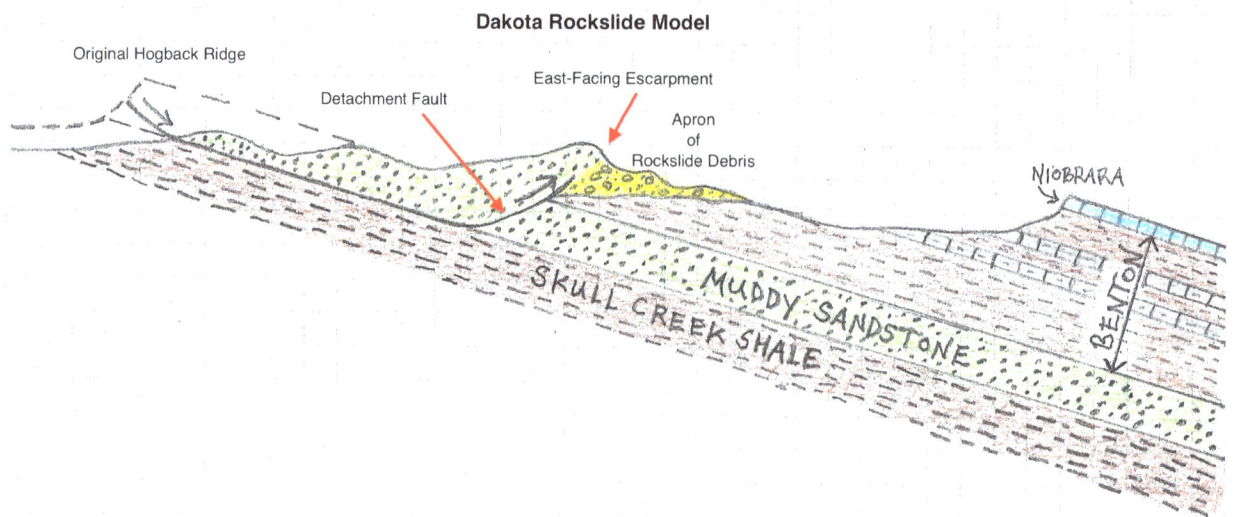

Fig. 56 Diagrammatic illustration of a translational rockslide— common along the Dakota hogback from Fort Collins to Soapstone Prairie. (Modified from Braddock, 1978)

Fig. 57a Aerial photo looking north toward Milner Mountain along the axis of the same named anticline. Uplift of the basement rocks composing Milner Mountain caused the arching of the sedimentary rock layers above. Subsequent erosion has removed the crest of the anticline leaving the dipping limbs of the fold exposed on both sides of the Lykins valley.

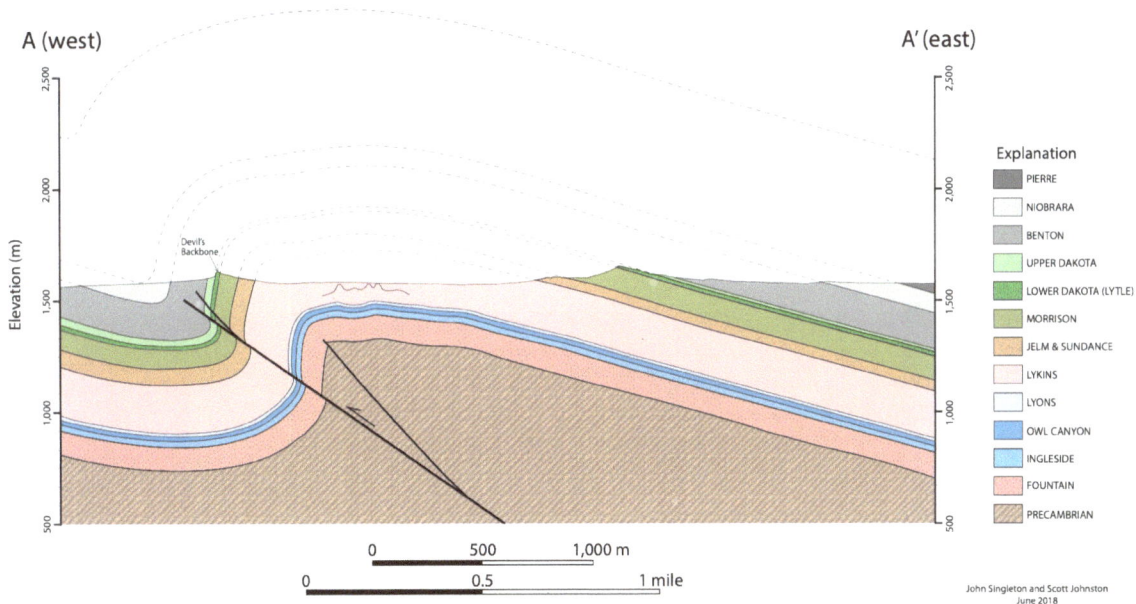

Fig. 57b Cross section across Devil's Backbone Open Space. Note that the eroded Milner Mountain Anticline in the middle of the diagram is now topographically inverted with the low-lying Lykins valley occupying the crestal part of the anticline. Steeply dipping, weather resistant sandstone and conglomerate of the Dakota Group (Lyle Formation) forms the "backbone" on the west limb of the anticline. (Courtesy of John Singleton, Colorado State University)

77

Fig. 58a The Muddy Sandstone located on the south side of Spring Creek Dam. A small pull-off area on W. County Road 38E allows for close examination. The Muddy is split into two sub-members: the brown weathered Horsetooth above, and the white colored Fort Collins below.

Fig. 58b Exposure of the Skull Creek Shale along Centennial Drive on the south side of Dixon Dam on Horsetooth Reservoir.

Fig. 58c An outcrop of the Plainview Sandstone made up of sugary white quartz grains interpreted as an ancient sand beach deposit. Located on Rimrock Trail near the connection with Coyote Ridge Trail.

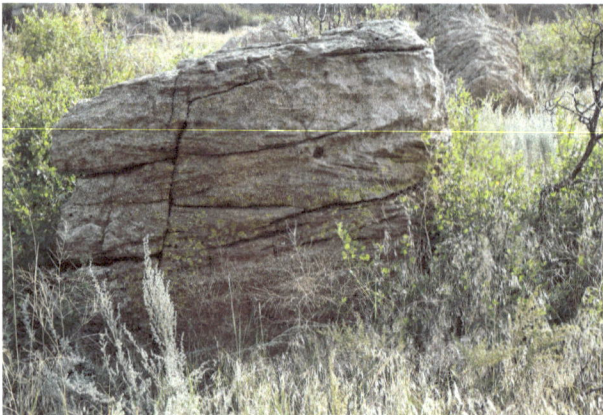

Fig. 58d Cross-bedded pebble conglomerate of the Lytle Fm. This block toppled down from the ridge above and sits next to the Rimrock Trail at Devil's Backbone Open Space.

BENTON GROUP

AGE: 98—90 million years ago (Late Cretaceous)

THICKNESS: 485—570 feet

DEPOSITIONAL ENVIRONMENT: Marine deposits of the Western Interior Seaway

LITHOLOGY: Shale, chalky limestone, marlstone, and sandstone

WHERE ARE GOOD EXPOSURES OF THE BENTON GROUP?

Coyote Ridge:
    Coyote Ridge Trail - On the west slope of the Fort Hayes Limestone ridge there is an inconspicuous outcropping of about a 6 inch bed of Codell Sandstone.
    Ridge to Ridge Trail - The best outcropping of Benton rocks is unfortunately off trail, but can be found about halfway to the Prairie Ridge OS boundary, east of the the trail.
Pine Ridge:
    Valley to Ridge Trail - Connector trail exposes Benton limestone and Codell Sandstone beneath Fort Hayes Limestone at top of ridge.

HOW DO I IDENTIFY THE BENTON GROUP?

- Dominantly shale with interlayers of limestone
- Forms a valley between the Dakota hogback and the low-relief Fort Hayes Limestone ridge
- Dixon Reservoir, College Lake and Claymore Lake lay within the Benton valley
- Outcrops are rare due to vegetative cover
- Where exposed, the rocks are mainly a mix of dark gray shale and thin limestone beds

---

TOPICS OF INTEREST:

Peak Sea Level

All the Cretaceous age rocks within the northern Front Range foothills (Dakota through Pierre) were deposited within or adjacent to the ancient Western Interior Seaway. Sea levels at this time were over 600 feet higher than they are today, and peaked at over 800 feet higher during the period of Benton Group deposition.

What caused this Seaway to flood the continent? Magmatism was likely the trigger. Geologists have discovered that sea-floor spreading rates at mid-ocean ridges were as much as four times faster than they are today. Also, very large flood basalts were erupted on the ocean floor during this period. The consequence of the increased magmatism was swelling of mid-ocean ridges and the building of large subsea volcanic plateaus that effectively decreased the volume of the ocean basin, forcing the seas to flood low-lying continental areas.

During this same period of elevated sea levels, the west coast of North America was the site of frequent continental collision resulting in fold-thrust mountains that migrated as far east as central Utah (the Sevier Orogeny). The weight of these stacked thrust sheets caused the load-bearing crust and lithosphere to flex downward creating a low-lying basin adjacent to the thrust belt mountains, it is this area, known as a **foreland basin**, that became flooded by the Western Interior Seaway.

By observing the succession of rock types, geologists have been able to discern five major cycles of fluctuating sea levels from the Western Interior Seaway rock record. A cycle is made up of a period of rising sea level (transgression) followed by a period of falling sea (regression). An idealized rock succession would start with a beach or shallow marine sandstone, followed by a deeper marine shale, then possibly an even deeper marine limestone or **chalk**, before reversing this sequence as sea level falls. The chalky limestones of the Greenhorn Fm (Benton) and Niobrara Fm were deposited at peak sea levels during their respective cycles.

Climate, Black Shales and Hydrocarbons

The increased magmatism noted above also affected earth's atmosphere creating a Cretaceous hothouse climate. Global average surface temperatures are thought to have been 10 degrees Celsius higher than they are today due largely to volcanically emitted CO_2 gases. Higher temperatures in turn caused earth's hydrologic cycle to move at faster rates. A wet climate quickens the weathering and erosion of rocks, ultimately feeding river borne minerals and nutrients to the sea, escalating marine organic productivity.

As sea temperatures rose, the dissolved oxygen content within the oceans decreased and became so severe during the time of Benton deposition that it is referred to as an Ocean Anoxic Event. Aerobic microorganisms that would normally live on the sea bottom and feed on the dead organic matter could no longer exist. The global effect was an increase in black shale deposition that reflects the accumulation and preservation of dead organic matter. Preserved and concentrated organic matter, when buried and cooked over geologic time, is the source of oil and gas. In fact the majority of hydrocarbons produced worldwide are from Cretaceous rocks.

In the Denver Basin, over 90% of the oil and gas produced is from Cretaceous rocks. The major black shale source rocks include: Skull Creek Shale (Dakota), Mowry Shale (Benton), Graneros Shale (Benton), Niobrara Marlstone, and the Sharon Springs Shale (Pierre). The hydrocarbons created within these organic rich source rocks migrated into adjacent porous sandstone and limestone reservoirs. In the Denver Basin the chief reservoir targets pursued by oil and gas companies are the Niobrara Chalks, the Muddy Sandstone (Dakota), and the Codell Sandstone (Benton).

Benton Group Oil and Gas

The Codell sandstone member of the Carlile Formation has been a historic target for oil and gas drilling within the Denver Basin since 1981. The majority of production comes from the giant Wattenberg field in Weld County. The Codell is a fine-grained low **permeability** sandstone that has always required stimulation (fracturing) to establish commercial flow rates. Starting around 2010, modern horizontal drilling and multi-stage fracture techniques have been successfully applied toward developing new resources within the Codell.

Within the last five years, the Lincoln and Bridgecreek Limestone members of the Greenhorn Formation have also become exploration targets using these same techniques. Like the Niobrara, the chalky Greenhorn limestones have sufficient **porosity** to be potential reservoirs and are adjacent to mature hydrocarbon generating **source rocks** within the Graneros and Hartland Shales.

SCHEMATIC ROCK COLUMN
BENTON GROUP

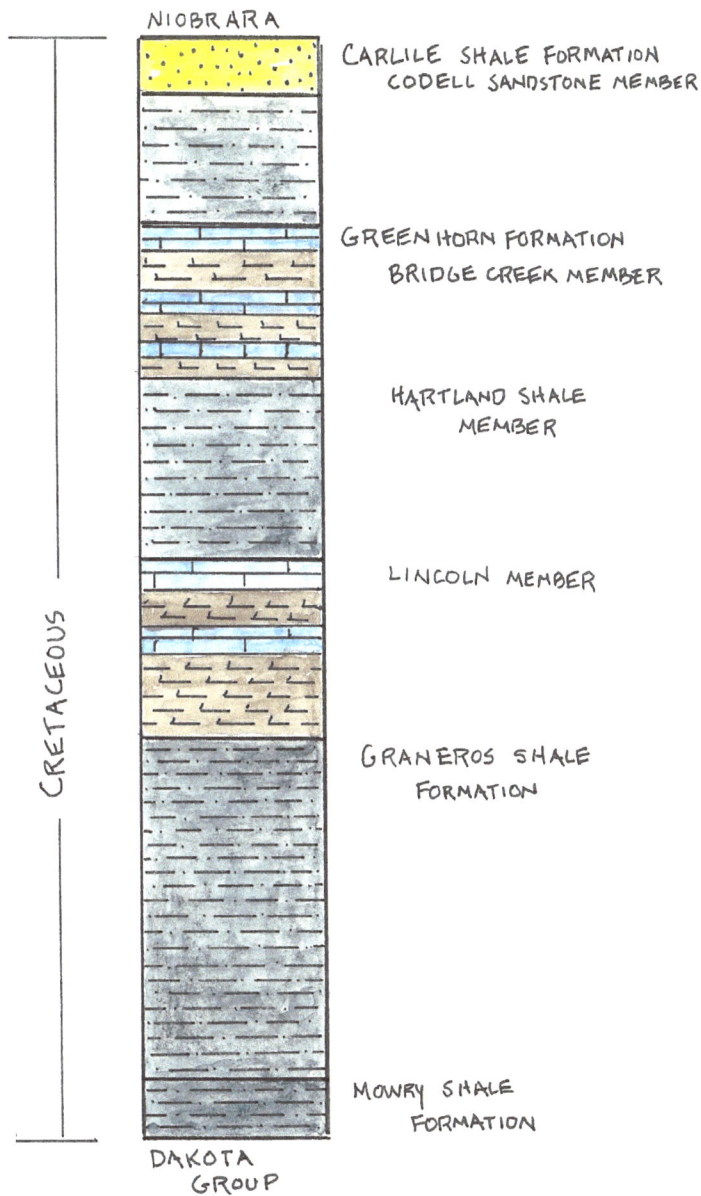

NIOBRARA

CARLILE SHALE FORMATION
CODELL SANDSTONE MEMBER

GREENHORN FORMATION
BRIDGE CREEK MEMBER

HARTLAND SHALE
MEMBER

LINCOLN MEMBER

GRANEROS SHALE
FORMATION

MOWRY SHALE
FORMATION

CRETACEOUS

DAKOTA
GROUP

Fig. 59 The Benton Group rocks record a cycle of rising and falling sea level referred to as the Greenhorn Cycle. At the height of sea level rise, which marked the widest extent of the Western Interior Seaway, interbedded chalky limestones and marlstones were deposited in the deeper part of the basin.

81

CRETACEOUS STAGES	ROCK UNIT	SEA-LEVEL CYCLES

Fig. 60
Sea level cycles and corresponding rock record of the Western Interior Seaway. Rock types tend to track rising and falling sea levels. The blue bars indicate periods of chalk deposition during the Greenhorn and Niobrara peak sea levels.
(Modified from Pang, 1995)

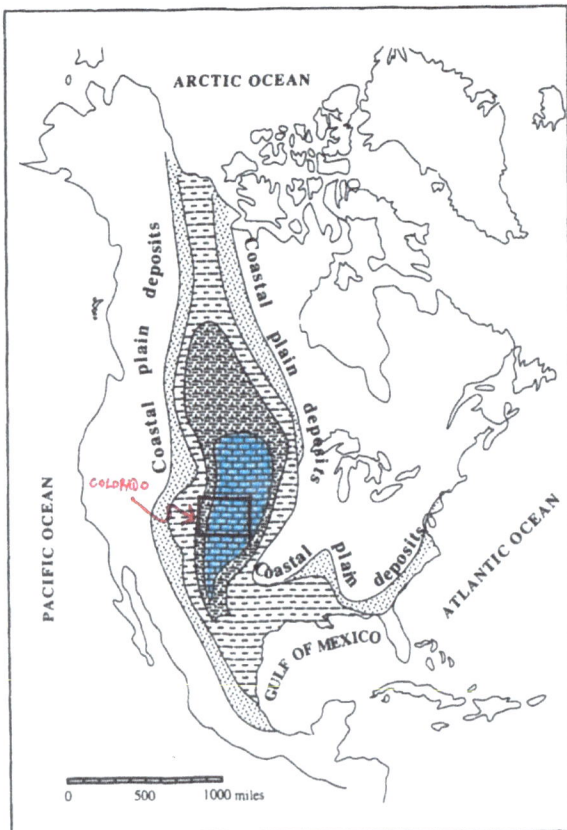

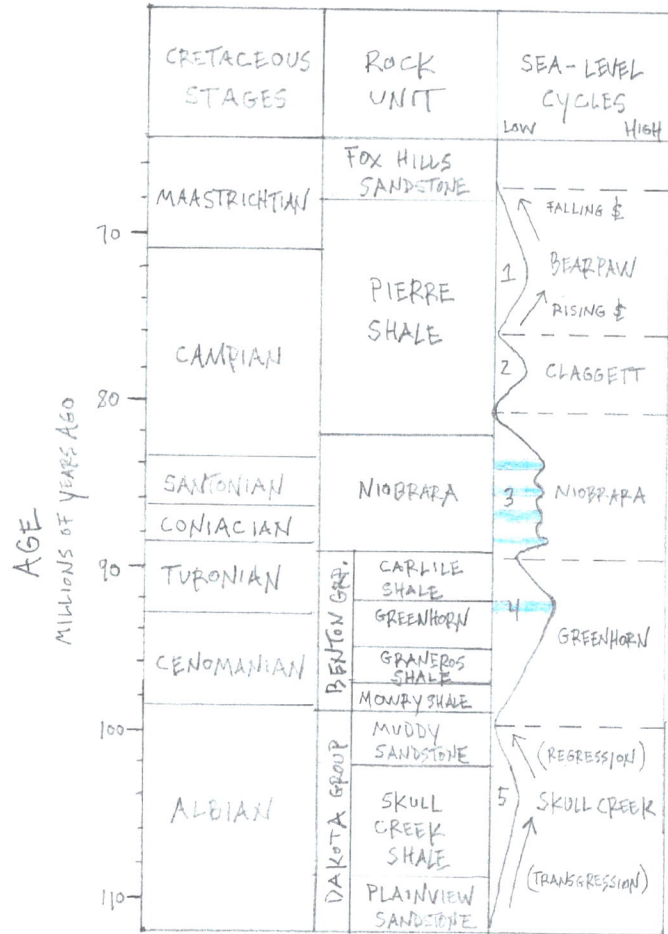

Fig. 61
A snapshot in geologic time of sedimentary rock types deposited at the peak of the Greenhorn cycle. At this time in Colorado, chalk and marl deposition dominated. As sea level dropped, the Carlile shale became dominant, capped by the regressive Codell sandstone. (Modified from Pang, 1995)

Fig. 62a Rare exposure of shale and limestone beds of the Benton Group Greenhorn Formation (gray colored rocks towards the middle of the photo). View is looking south from the top of the Niobrara Fort Hayes Limestone ridge in Coyote Ridge Natural Area.

Fig. 62b On the trail segment connecting the Valley Loop and Ridge Trails at Pine Ridge Natural Area, staircase ledges expose the contact between the Niobrara Formation and Benton Group rocks. The upper white colored ledges are the Fort Hayes Limestone Member of the Niobrara Fm and the lower buff colored ledge is the Codell Sandstone Member of the Carlile Shale Formation—part of the Benton Group.

Fig. 62c This outcrop—unfortunately not accessible by trail—is located on the north side of Pine Ridge NA. It is best seen from Dixon Canyon Road. The upper half of the cliff face is white colored Fort Hayes Limestone (member of the Niobrara Fm) that is in contact with the Carlile Shale member of the Benton Group below.

Fig. 62d Close-up view of the figure above. The thin, brown weathered layer is the Codell Sandstone—part of the Carlile Shale. The Codell Sandstone is a major oil and gas producer in the Denver Basin.

NIOBRARA FORMATION

AGE: 82—89 million years ago (Late Cretaceous)

THICKNESS: 350 feet

DEPOSITIONAL ENVIRONMENT: Deep-water marine of the Western Interior Seaway

LITHOLOGY: Limestone, marlstone, and chalk

WHERE ARE GOOD EXPOSURES OF THE NIOBRARA?

Coyote Ridge:
  Coyote Ridge Trail - The first low-relief rise on the trail is formed by the A Chalk— the youngest part of the Smoky Hill Member of the Niobrara Fm. The trail continues over the Niobrara until reaching the oldest member known as the Fort Hayes Limestone which forms a prominent low relief ridge. The trail follows this ridge north until descending into the valley underlain by the Benton Group rocks.

Pine Ridge:
  Ridge Trail - This trail traverses on top of the fossiliferous Fort Hayes Limestone.

Reservoir Ridge:
  Michaud Spur Trail - The trailhead parking lot and beginning part of the trail lay on top of the low-relief ridge formed by the Fort Hayes Limestone.

Boettcher Quarry:
  Overland Trail Road; north of US 287—exposure of the Smoky Hill Member within the quarry.

HOW DO I IDENTIFY THE NIOBRARA?

- The A Chalk typically weathers to a dusty yellow color
- Otherwise, the Smoky Hill Member chalks and Fort Hayes Limestone are light gray in color
- Variously abundant fossils, but otherwise a **microcrystalline** texture
- Clinking two plates of the rock together will produce a brittle sound, like ceramic plates

---

TOPICS OF INTEREST:

Cretaceous Chalk

The Cretaceous (latin Creta for chalk) Period is named for the ubiquitous chalks which were deposited worldwide during this geologic time period. A well-known example being the White Cliffs of Dover on the SE coast of England.

To a geologist a chalk is a very pure fine grained skeletal limestone (calcium carbonate) made from the remains of microscopic plants (algae) and animals (forams) that lived floating in the ocean. A particular algae known as a **coccolithophore** produces an exoskeleton comprised of a number of beautifully sculpted overlapping plates made of calcium carbonate that look like a great coat of armor. The plates, called coccoliths, accumulate on the ocean floor and when later buried by further sedimentation are turned into the special form of limestone we call chalk.

The impressive chalk deposits of the Cretaceous are causally tied to a period of very active plate tectonics that caused a global rise in sea level (forming the Western Interior Seaway), and created a much warmer climate that drove increased productivity of coccolithophores.

In our foothills outcrops, the Niobrara chalks are neither white nor as soft as the Cliffs of Dover chalks. Deposition in this part of the Western Interior Seaway was not as pure in the fraction of skeletal remains (eg, coccoliths). The interbedded marlstones are rocks that by definition have a significant component of quartz silt and clay mixed in with the calcium carbonate content.

Fossils

Besides the microscopic chalk-forming coccoliths mentioned above, the Niobrara is also rich in fossils we can see with the naked eye. One of the most common is of an extinct genus of clams known as **inoceramous**. Over 70 species of inoceramous have been identified, and they range in size from about 1 inch to giants over 4 feet in diameter. Fossils of the larger species have been found with attached colonies of smaller oysters that grew on their shells. And even more remarkably, fossil fish and pearls have been found on the inside of some inoceramous fossils. Over the course of a 34 million year history inoceramous evolved rapidly, allowing geologists to use different species as index fossils for rock correlation and dating the rock zones in which they are found.

Quarrying

In 1927, the Ideal Cement Company began strip mining the Niobrara to support its Laporte, Colorado cement plant. The focus of quarrying was the Fort Hayes Limestone which was found to have the more pure calcium carbonate content (less clays and silts). The strip mining occurred along the Fort Hayes Limestone ridge for a stretch of more than five miles. The loose tailings from the quarrying operation can be seen along US 287 north of the intersection with Overland Trail Road. The tailings were built up and shaped to appear at first glance like a hogback to blend in with the natural surroundings. The plant and quarry closed in 2002.

The Fort Hayes Limestone is currently being quarried to support the Cemex plant just outside of Lyons on Route 66. The giant quarry can be seen from trails on Rabbit Mountain Open Space in Boulder County.

Oil & Gas

The Niobrara has been a drilling target in the Denver Basin for many decades. Beginning around 2010 however, armed with new technologies for drilling and completing horizontal wells, oil companies created a huge boom in activity targeting the Niobrara in Weld County, Colorado. The targeted oil reservoirs are the chalk units of the Smoky Hill member, especially the prolific B Chalk. Analysis has shown that the adjacent marlstone units have high organic carbon concentrations and are the source rocks for the hydrocarbons that have migrated into the porous chalks. As of December 2019, total Niobrara oil production was 700,000 barrels/day.

In Larimer County, which sits on the western edge of the Denver Basin, the Niobrara is too shallow to be a viable oil reservoir. Lower temperature and pressure is not adequate for the organic source rocks to 'cook' oil, and not enough pressure resides within the potential reservoirs to produce oil at economic rates.

SCHEMATIC ROCK COLUMN
NIOBRARA FORMATION

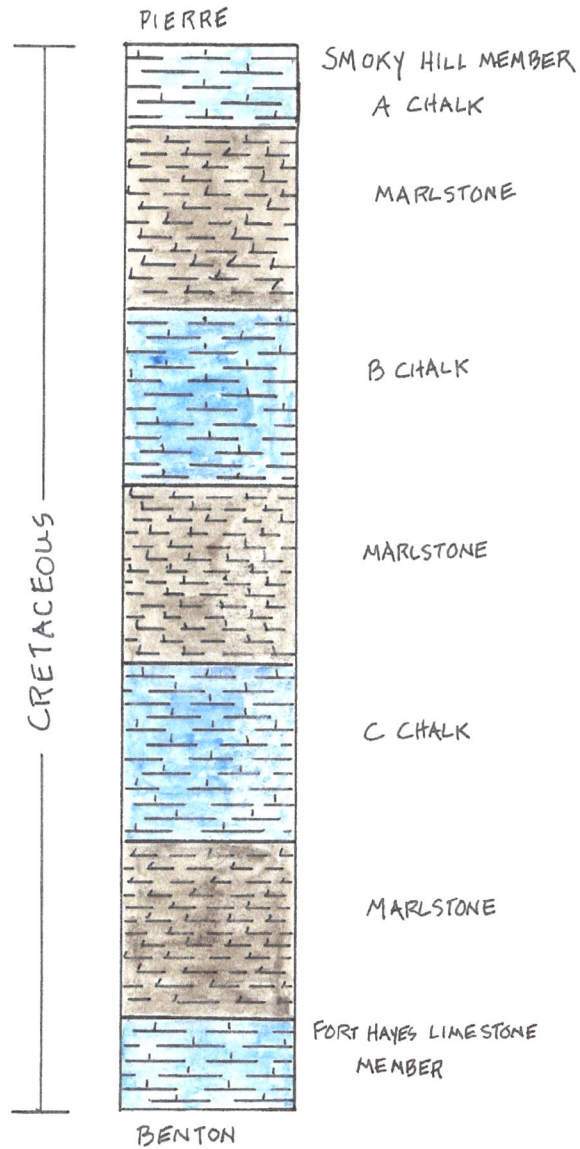

PIERRE

SMOKY HILL MEMBER
A CHALK

MARLSTONE

B CHALK

MARLSTONE

C CHALK

MARLSTONE

FORT HAYES LIMESTONE MEMBER

CRETACEOUS

BENTON

Fig. 63 The Niobrara chalks and marlstones were deposited within the deepest part of Western Interior Seaway during an extended 5 million year period of high sea level.

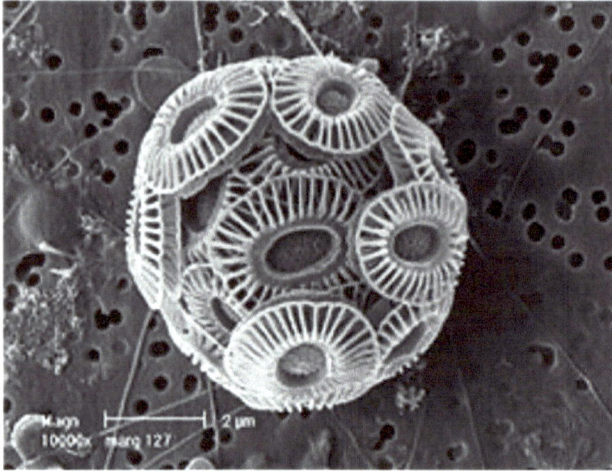

Fig. 64 The hubcap shaped coccoliths are a primary constituent of the Niobrara chalk layers. (web.colby.edu)

A coccolithophore highly magnified

Fig. 65a An example inoceramus fossil in the Fort Hayes Limestone. Coyote Ridge Natural Area.

Fig. 65b Exposure of the A chalk located at the first rise on the Coyote Ridge Trail.

Fig. 66 View looking north along the Boettcher Quarry that was mined for the high calcium carbonate content of the Fort Hayes limestone used in the manufacture of concrete. Bordering the mine pit on the left are the piled up tailings of overburden that were removed to get access to the Fort Hayes. US 287 sits just west of this tailings mound. On the right bank of the quarry is the Smokey Hill Member of the Niobrara Fm.

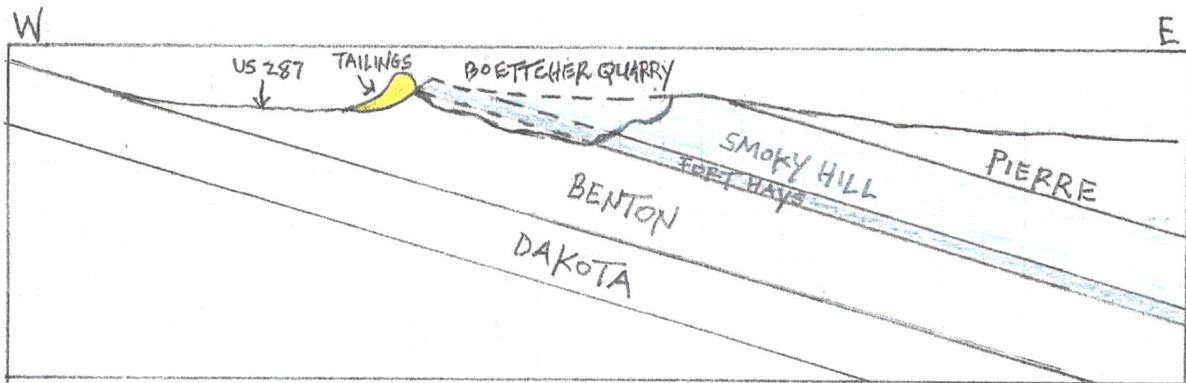

Fig. 67 Schematic cross section illustrating the underlaying geologic structure along the Boettcher Quarry. Because all the beds dip to the east, the cost of quarrying became uneconomic as the Fort Hayes sunk deeper into the subsurface. As a result the quarrying operation followed the formerly outcropping ridge of the Fort Hayes north for over 5 miles. The hachured area represents rock that was removed during the quarrying operation.

Fig. 68a West facing wall of the Boettcher Quarry exposing the tan colored 'A' Chalk overlying darker colored marlstone.

Fig. 68b Photo taken from Maxwell NA looking south toward Pine Ridge NA. In the mid-distance is the low-relief ridge commonly formed by the Fort Hayes Limestone.

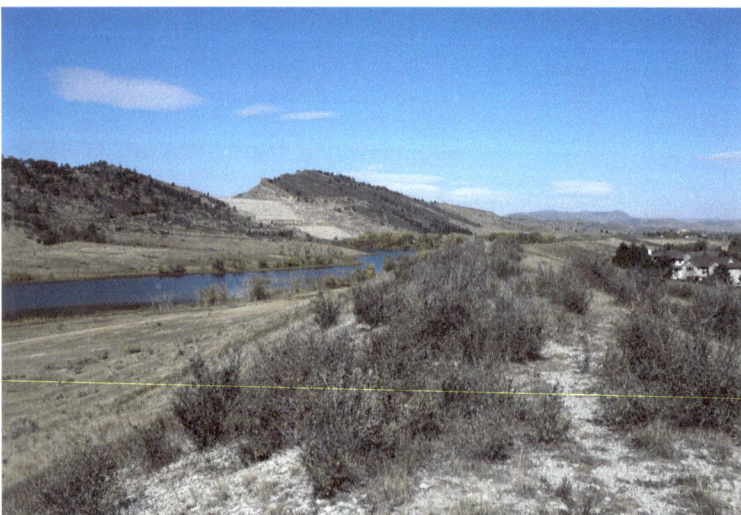

Fig. 68c Pine Ridge Natural Area. The gray rocks in the foreground lay on the low-relief ridge formed by the Fort Hayes Limestone. Dixon Reservoir on the left sits in the shallow valley underlain by rocks of the Benton Group. The housing on the right is part of a development built on top of the Smoky Hill Member of the Niobrara.

PIERRE SHALE

AGE: 72—82 million years ago (Late Cretaceous)

THICKNESS: 6,800 feet

DEPOSITIONAL ENVIRONMENT: Marine deposits of the Western Interior Seaway

LITHOLOGY: Shale, siltstone, and sandstone

WHERE ARE GOOD EXPOSURES OF THE PIERRE SHALE?

The Pierre (pronounced pier, as in boat pier) Shale is the thickest sedimentary rock unit of the foothills and is the underlying bedrock extending from the base of the foothills, to beneath Fort Collins, and east to Windsor. And although the Pierre was tilted up by growth of the Rocky Mountains along with the other formations of the foothills hogback belt, erosion has mostly beveled the formation so that it now forms the western edge of the Great Plains.

Coyote Ridge:
    Coyote Ridge Trail - Exposure of black flaky shale of the Sharon Springs Member on barren ground just before the contact with the Niobrara.

Soapstone Prairie:
    Pronghorn Loop Trail - Exposure of the Mitten Black Shale Member along the banks of the Wire Draw drainage just east of the entry gate.

Colina Mariposa NA Trail - Many fossil fragments and cannonball **concretions** can be seen weathering out of the Rocky Ridge Sandstone Member.

HOW DO I IDENTIFY THE PIERRE SHALE?

- Olive gray to black; flaky, fissile shale; platy bedding
- Common concretions
- Exposures are rare. Covered by vegetation, alluvial sediment, and urban development

---

TOPICS OF INTEREST:

Oil and Gas

Colorado's oldest oilfield, discovered in 1862, is the Florence Field located west of Pueblo. It produces from fractured Pierre Shale. In the 1970s production was also established from the Hygiene and Terry Sandstone Members of the Pierre Shale in Weld County. And of historic importance, in Larimer County there are two active oil fields, now approaching almost a century old, that don't produce from the Pierre Shale but were discovered by surface mapping of Pierre Shale sandstone outcrops. The Fort Collins and Wellington oilfields were both discovered circa 1924 and lie just a few miles from their respective town centers. Mapping out the geometry of the exposed Pierre sandstone members indicated that a possible enclosed oil trap (anticline) existed beneath the surface. Subsequent drilling led to the discovery of oil in the Muddy (Dakota) Sandstone about 4000 feet below the surface. This consequential discovery led to expanding Muddy Sandstone production throughout the Denver Basin and became one of its most economic reservoir targets.

Geologic Hazard

Perhaps surprisingly, the most significant geologic hazard in Colorado is not from catastrophic floods or landslides, but rather due to property damage from swelling soils. The swelling is due to a clay mineral known as montmorillonite which has the property of being able to absorb water molecules within its structure. In a soil rich in montmorillonite, the volume expansion can range as high as 20-50% creating swell pressures of up to 20,000 pounds of force per square foot. This is a big problem, considering that structural engineers are concerned with expansions of only 3% as potentially damaging. The primary source of the montmorillonite is from **weathered volcanic ash** deposits, also known as **bentonite**. Parts of the Pierre Shale have layers of volcanic ash derived from volcanic eruptions that were laid down as wind-borne deposits and settled on the floor of the Western Interior Sea. With 80% of Colorado's population residing on the eastern margin of the Front Range and much of the development on top of Pierre Shale **bedrock**, there are many pockets of exposure to this problem. Structures built across tilted bedrock layers of the Pierre Shale can potentially cross over a layer of bentonite. This can lead to **differential pressure** across a structure's foundation and may cause structural damage. The problem has been particularly widespread in Jefferson County southwest of Denver.

Fort Collins Type Sections for Pierre Sandstones

Most of the rock formations outcropping in the foothills were first described and named after type localities within the foothill belt running from Boulder to Livermore. The Pierre Shale Formation is an exception, being first described and named near Fort Pierre in South Dakota. But the sandstone members comprising the middle part of the Pierre Shale were all described and named for exposures near Fort Collins. In 1924, geologist M. W. Ball named the Terry, Rocky Ridge and Richards Sandstones because outcrops of each of these sandstone units were exposed near the banks of the respective water reservoirs just a few miles northeast of Fort Collins. He also named the Larimer Sandstone member due to exposures of this unit within the Larimer County Canal about four miles north of Fort Collins. All except the Richards Sandstone are rich with fossil-bearing concretions.

Concretions and Fossil Ridge

A conspicuous feature of the Pierre Shale is the common occurrence of concretions. Concretions are relatively hard masses of rock embedded within softer surrounding rock that form as a result of focused **cementation**. The cements, which are dissolved minerals such as limestone, will often precipitate around a fossil. Concretions are thought to form early in the burial history of a sediment, before fully becoming a rock, and while still in contact with groundwater. The size of reported concretions within the Pierre Shale can be on the scale of inches to greater than 10 feet in diameter.

Since 1879, and over a period of about 50 years, geologists have studied a ridge of outcropping sandstone that extends for about 4 miles south of Trilby Road and between US 287 and Shields Street. To geologists it is known as Fossil Ridge due to the treasure trove of fossils found on the western side of the ridge where the Rocky Ridge Sandstone member is exposed. Many species of marine invertebrate fossils were obtained from the site, typically within 'cannonball' concretions. **Ammonites**, an extinct mollusk with a chambered external shell are abundantly present as fossil fragments at the site. The ridge is currently protected by Colina Mariposa Natural Area and Long View Farm Open Space.

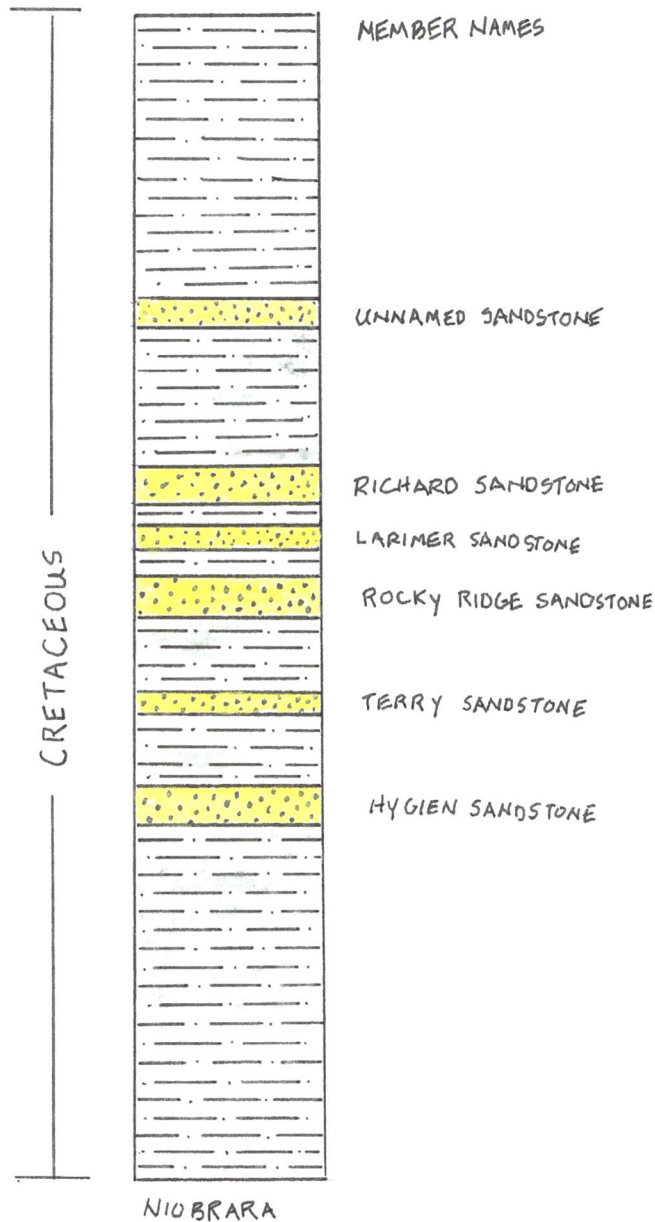

Fig. 69 The youngest and thickest formation in the northern Colorado Front Range foothills, the Pierre Shale is dominantly a shale with thin sandstone members (exaggerated thickness above) situated towards the middle of the formation. Deposition of the unit was all within the Western Interior Seaway.

Fig. 70a Concretions weathering out of the Rocky Ridge sandstone member of the Pierre Shale, Colina Mariposa Natural Area, Fort Collins, Colorado.

Fig. 70b Large ammonite recovered from the Pierre Shale. Colorado School of Mines Geology Museum.

Fig. 70c Black fissile shale flakes of the Pierre are exposed in a bald area about ten yards south of the Coyote Ridge Trail— just before the contact with the underlying Niobrara at the first rise in the trail.

Fig. 70d Olive and dark gray colored Pierre Shale exposed within an arroyo on the east side of the road just past the entry station at Soapstone Prairie Natural Area.

PART B

Creation of the Rocky Mountains—the Laramide Orogeny

View looking southwest from the Mahogany Trail in Soapstone Prairie toward Long's Peak on the left, and the Mummy Range on the right. The Rocky Mountains were uplifted from 70—40 mya. During this time all the sedimentary rock units discussed in Part A were arched upward and eroded off the top of the mountains, exposing the Precambrian core.

THE LARAMIDE OROGENY

The entire suite of rock units discussed in Part A were subject to the mountain building event geologists refer to as the Laramide Orogeny, the creator of today's Rocky Mountains. For the Precambrian rocks, the Laramide Orogeny was the third mountain-building event to have affected them. Recall that these rocks were formed during a time of **continental collision** nearly 1.8 billion years ago, in effect creating the state of Colorado. And again about 300 million years ago were subject to mountain building forces that formed the Ancestral Rocky Mountains, part of another continental collision that ultimately produced the supercontinent Pangaea. The overlying sedimentary rocks on the other hand, from the Fountain to Pierre Shale formations were deposited and layered horizontally at the earth's surface only to be arched upward, faulted and folded during the Laramide Orogeny roughly 70—40 million years ago. Subsequent erosion has exposed in wonderful landscape relief the tilted, folded and faulted rocks along the northern Colorado Front Range foothills.

Plate Tectonic Background

Plate tectonics, geology's grand unifying model explaining the cause of mountain belts, volcanoes, earthquakes and other geologic features was generally accepted with the support of much gathered data by the late 1960s. The earth's outer layer is now known to be composed of about a dozen major **lithospheric** plates that are driven by **convection currents** within the **asthenosphere** and move with respect to one another over earth's surface. There are two major types of plates: those that are capped with buoyant **continental crust** and those that are capped with denser **oceanic crust** which underlay the earth's ocean basins. When an oceanic plate converges with a continental plate its greater density will cause it to subduct beneath the continental plate and into the asthenosphere. The compressive forces that result from colliding plates result in major mountain chains near the plate boundaries. Also, the subducted plate of oceanic crust liberates water that has the effect of lowering the melting temperature of the asthenospheric mantle, creating igneous plutons and volcanoes adjacent to the subduction zone.

The Laramide Enigma

The Laramide Orogeny produced a distinctive chain of basement cored uplifts and adjacent basins extending from southern Montana to southern New Mexico. Why this style of mountain building was developed so far from the existing **convergent plate margin** (+/- 1000 miles to the west) was a puzzle that did not fit the model of plate tectonics. The older Sevier Orogeny (150—65 mya) produced an eastward propagating fold and thrust belt that terminates in central Utah. It is logically understood as a result of the horizontally directed compressive forces transmitted through the continental crust due to the convergence of the Farallon and North American plates. The style of deformation is considered 'thin-skinned' because the low angle thrust faults are mostly confined to the Paleozoic and Mesozoic sedimentary rocks near the earth's surface. To explain the Rocky Mountains geologists proposed that the subducting Farallon plate flattened beneath the overriding North American plate. Evidence suggests flattening was brought about by the subduction of a thick and buoyant oceanic plateau that entered the subduction zone near southern California about 80 million years ago. Using advanced seismic reflection technology, geophysicists have convincingly imaged relict fragments of the thickened Farallon plate beneath the central United States. In the new plate tectonic model, the Laramide Orogeny was produced by more vertically inclined forces above and adjacent to the hinge line where the flattened Farallon plate steepened and subducted into the asthenosphere. The foreland basin was thus broken up into a series of 'thick-skinned' basement cored uplifts and adjacent basins that characterize the Rocky Mountains.

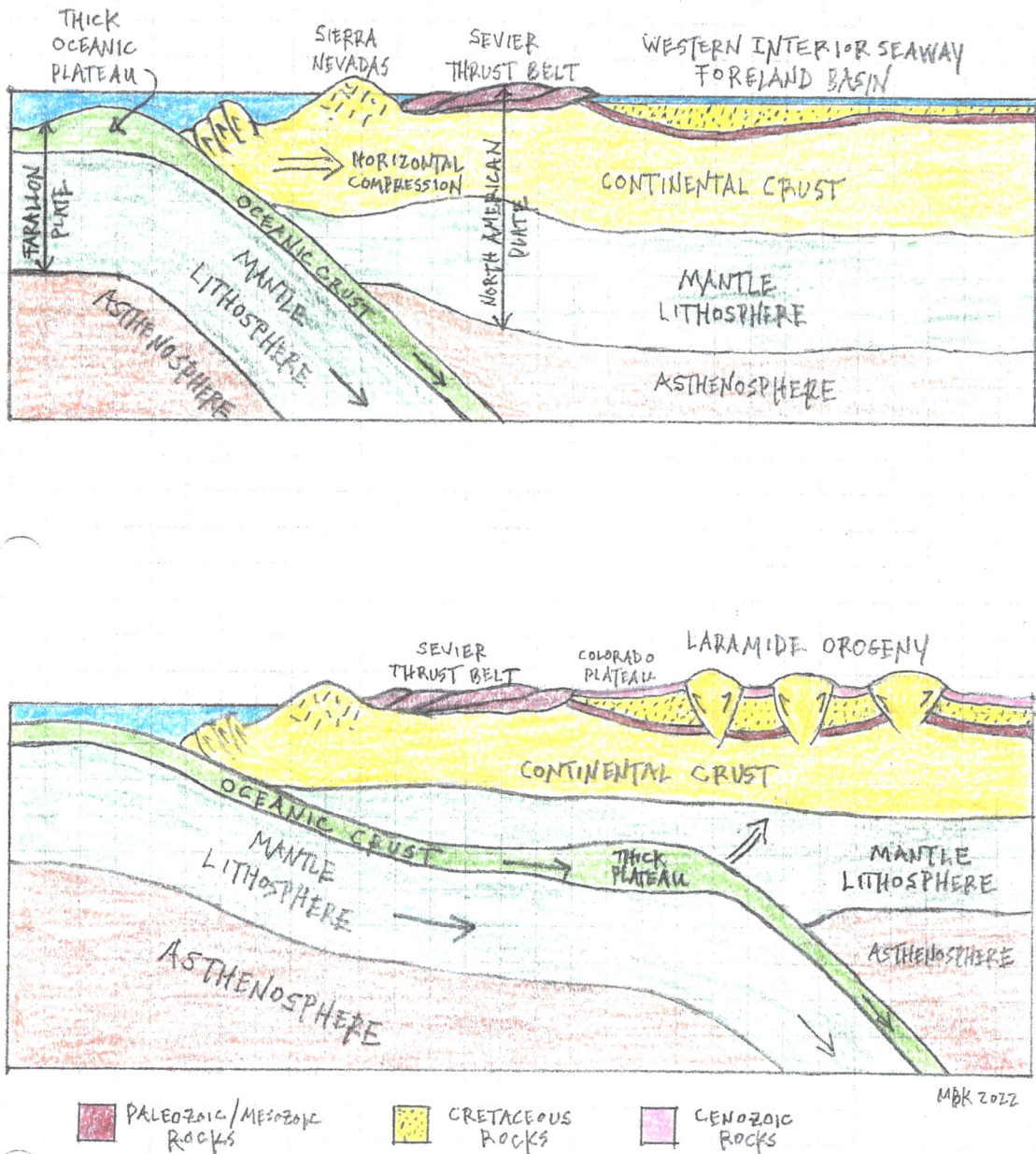

Fig. 71 Schematic diagrams showing the plate tectonic settings during the Sevier and Laramide Orogenies. **Top Diagram**: Beginning in the late Jurassic Period, horizontal compressive stresses associated with the convergence of the oceanic Farallon plate with the North American continental plate created the Sevier thrust belt and adjacent foreland basin. During the Cretaceous Period, the Western Interior Seaway spilled into the foreland basin where thick sedimentary deposits became the rock units we now recognize as the Dakota through the Pierre Shale. **Bottom Diagram:** During the Laramide Orogeny the subducting Farallon Plate flattened and pushed further east beneath the overriding North American plate. More vertically inclined compressive stress developed at the leading edge of the flattened Farallon plate producing the breakup of the foreland basin into basement cored uplifts and adjacent basins. (Modified from Blakey & Ranney 2018, and Gutscher 2018)

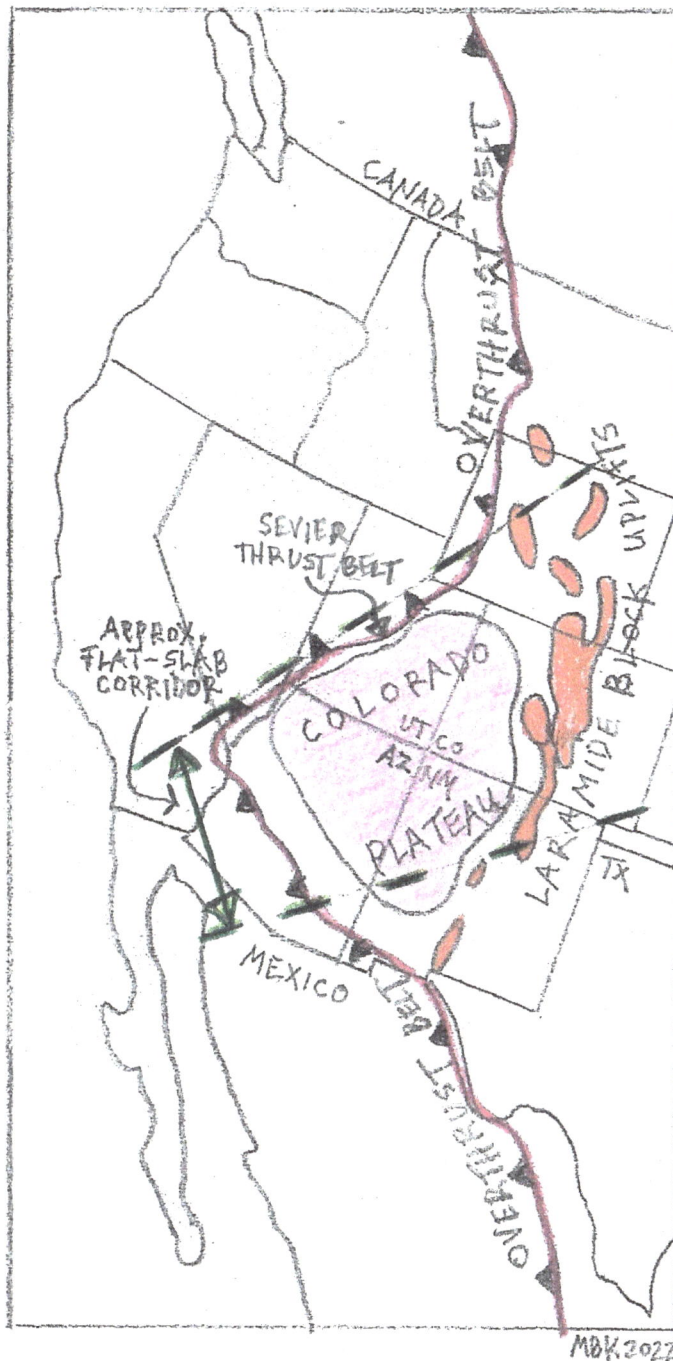

Fig. 72 The thick-skinned basement cored uplifts created by the Laramide Orogeny (orange color) form mountain ranges spanning from southern Montana to southern New Mexico. The eastern limit of the thin-skinned Sevier thrust belt runs through central Utah along the western boundary of the relatively undeformed Colorado Plateau. The area between the dashed green lines represents the corridor along which the flat-slab part of the Farallon Plate passed.
(Modified from English and Johnston, 2004)

PART C

Post Laramide Rock Units

at

Soapstone Prairie Natural Area

Aerial photo looking west along the Lindenmeier Valley. The ridge of rocks along the south margin of the valley are an erosional outlier. Erosion continues to strip off the relatively young, flat-lying rock units on display at Soapstone Prairie— rocks that once covered all the natural areas discussed in this guide—exposing the tilted and folded rocks below that were deformed during the Laramide Orogeny. In the process, the southern margin of the High Plains, locally defined here by the northern margin of the Lindenmeier Valley, continues to recede northward toward Wyoming.

Soapstone Prairie is uniquely different from the other natural areas discussed in this guide due to its location—straddling the Piedmont-High Plains boundary—and the much younger age of the rocks—deposited after the Laramide Orogeny. While Soapstone is not technically part of the Front Range foothills, its western boundary is shared with Red Mountain Open Space.

What makes Soapstone geologically special is that the rocks preserved here are a record of geologic history that was removed by erosion from the other natural areas. Whereas the Pierre Shale is the youngest rock unit in the northern Front Range foothills natural areas, it is the oldest rock unit at Soapstone Prairie.

TOPICS OF INTEREST:

Ogallala Aquifer

The Ogallala Formation records a second pulse of Rocky Mountain uplift that resulted in vigorous erosion of coarse grained rock fragments derived from the crystalline core of the mountains. This is documented by the constituent fragments of granite and metamorphic rocks, as well as large quartz and feldspar crystals. At Soapstone Prairie the Ogallala is the brown to reddish-brown caprock on the south-facing escarpment; it rests on the light colored Arikaree and White River Group rocks that form the slope below.

The coarse sandstone and conglomerate of the Ogallala is porous and permeable; that is the rocks are able to store and transmit fluids. In the subsurface the Ogallala is an immensely important aquifer that supports one of the nation's prime agricultural areas on the Great Plains —an area once referred to as the 'Great American Desert' by early explorers. The Ogallala underlies parts of eight states and holds the estimated equivalent of over 20,000 Horsetooth Reservoirs in water volume. As the nation's largest groundwater reservoir it enables production of about 30% of our irrigated agriculture, equal to $20 billion dollars' worth of annual food and fiber production.

Widespread industrial irrigation began in the 1950's. There are now over 200,000 irrigation wells tapping into the Ogallala. But withdrawal rates are 3 to as much as 15 times the recharge rates which have lowered the level of the aquifer by as much as 200 feet. Current withdrawal rates are not sustainable. Many communities, local and state governments are now coming to grips with this impending disaster and are looking for ways to mitigate the decline and ultimately find sustainable solutions.

Ignimbrite Flare-Up

By the middle Eocene about 40 million years ago the Rocky Mountains were no longer being driven upward by the compressive forces of the shallowly subducting Farallon plate. As the Farallon plate began to roll back into the mantle a period of intense volcanism developed across the west. Large volcanic fields in Nevada, Utah and the San Juan Mountains of Colorado were the sites of explosive volcanic events creating huge plumes of volcanic ash that were carried eastward for hundreds of miles. Many of the calderas associated with these ancient volcanic events are still preserved in the rock record with some having diameters of over 30 miles. These were indeed super-eruptions.

Ignimbrites are volcanic deposits that show vertical and lateral variation as a function of gravitational settling with distance from the eruption site. Fine ash is the last to settle out and tends to be found at the top of an ignimbrite deposit as well as far away from the eruption.

Both the White River Group and Arikaree Formation contain significant amounts of volcanic ash as both free-all deposits and reworked river deposits. At Soapstone Prairie, these two rock units make up the light colored slope beneath the Ogallala Formation. On a clear day you can see these conspicuous light colored volcano-clastic rocks from Fort Collins.

<u>Lindenmeier Folsom Site</u>

Lindenmeier is a major archaeological discovery made in 1924 within the east-west trending valley situated at the base of the high plains plateau and now part of Soapstone Prairie Natural Area. The discovery documents the presence of Paleo-Indians who regularly inhabited the site at the close of the Pleistocene epoch. Referred to as Folsom people by archeologists, the site contains stone tools, projectile points and associated animal bones. Charcoal from the same layer dates the time of occupation to about 12,300 years before present. The Smithsonian Institution excavated the site in detail from 1935—40.

Looking southwest about 1,000 feet from the Lindenmeier Overlook at Soapstone Prairie you can see the south wall of a gulch that has been cut into the Lindenmeier Valley sediments. Along that south wall and at the base of the gulch is a layer of brownish black sediment as much as two feet thick that rests above white colored White River Group bedrock flooring the bottom of the gulch. This is the cultural layer from which the main artifacts have been discovered and documented.

Since the time of Folsom people occupation, the valley has been filled in with about 12 feet of sediment covering and preserving the artifacts. Between the time of Folsom occupation and the arrival of white settlers, the geological process of **headward erosion** and **stream capture** has exposed the cultural layer within the gulch that formed and is still working its way north and west up the valley. Directly south of the Lindenmeier Overlook is a north trending drainage that eroded its way north across the east-west trending drainage divide on the south side of Lindenmeier valley and captured the east flowing drainage within the valley. The gulch was created by down-cutting across the steeper gradient of the drainage divide.

<u>The Great Exhumation</u>

Soapstone Prairie straddles the break in slope between the Colorado Piedmont and the High Plains and as such is a good place to contemplate and visualize the work of erosion. Until about five million years ago, the volcanic ash laden deposits of the White River and Arikaree rocks, together with the overlaying Ogallala, formed a broad apron of sediment along the Front Range that essentially buried the mountain topography and formed a somewhat smooth and gently eastward sloping surface away from the mountain front. Beginning about 5 million years ago and continuing today, the geologic process changed from one of deposition to erosion. The Colorado Piedmont, extending from Soapstone to near Denver is a broad valley carved out of the High Plains by the erosion of more than 2,000 feet of sedimentary rock including the Ogallala, Arikaree and White River Group rocks. As a result cities like Fort Collins and Loveland sit 2,000 feet lower in elevation than the high plateau at Soapstone Prairie.

The High Plains rocks preserved at Soapstone are part of an erosional remnant that is still being attacked by tributaries of the North and South Platte River drainages. Hiking through Soapstone you will notice the gullies and gulches notched into the plateau, or simply looking at the corrugated contours on a topographic map of the area gives a good sense that geologically speaking this erosional remnant will not last very long. In the meantime, this erosional remnant, often referred to as the 'gangplank', forms a low gradient transportation ramp connecting the Great Plains to the mountain front and was conveniently used by the builders of the transcontinental railroad and Interstate 80.

Fig. 73 View looking NW toward the southern exposure of rocks most commonly associated with Soapstone Prairie. In the middle ground are the greenish gray rocks of the White River Group. Rocks of the light brown colored Arikaree Formation make up the lower slope of the ridge, while the darker brown caprock is the Ogallala Formation.

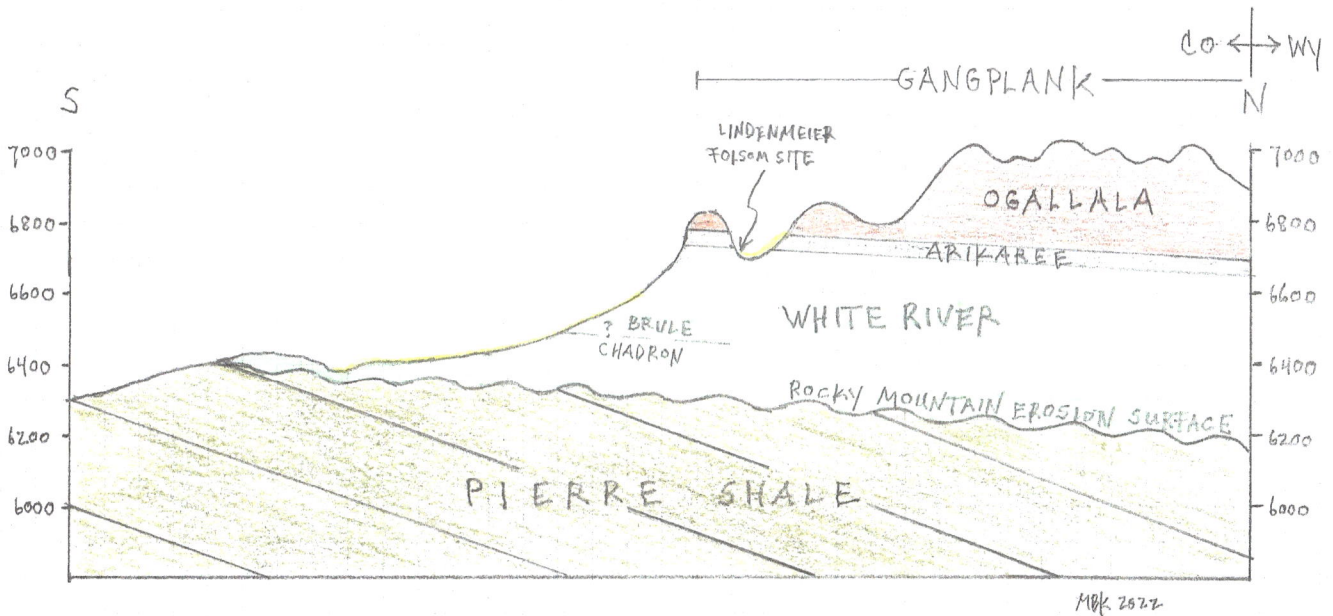

Fig. 74 North-south oriented geologic cross section showing rock units and topography at Soapstone Prairie. The same rock units also compose the Pawnee Buttes in eastern Weld County, Colorado.

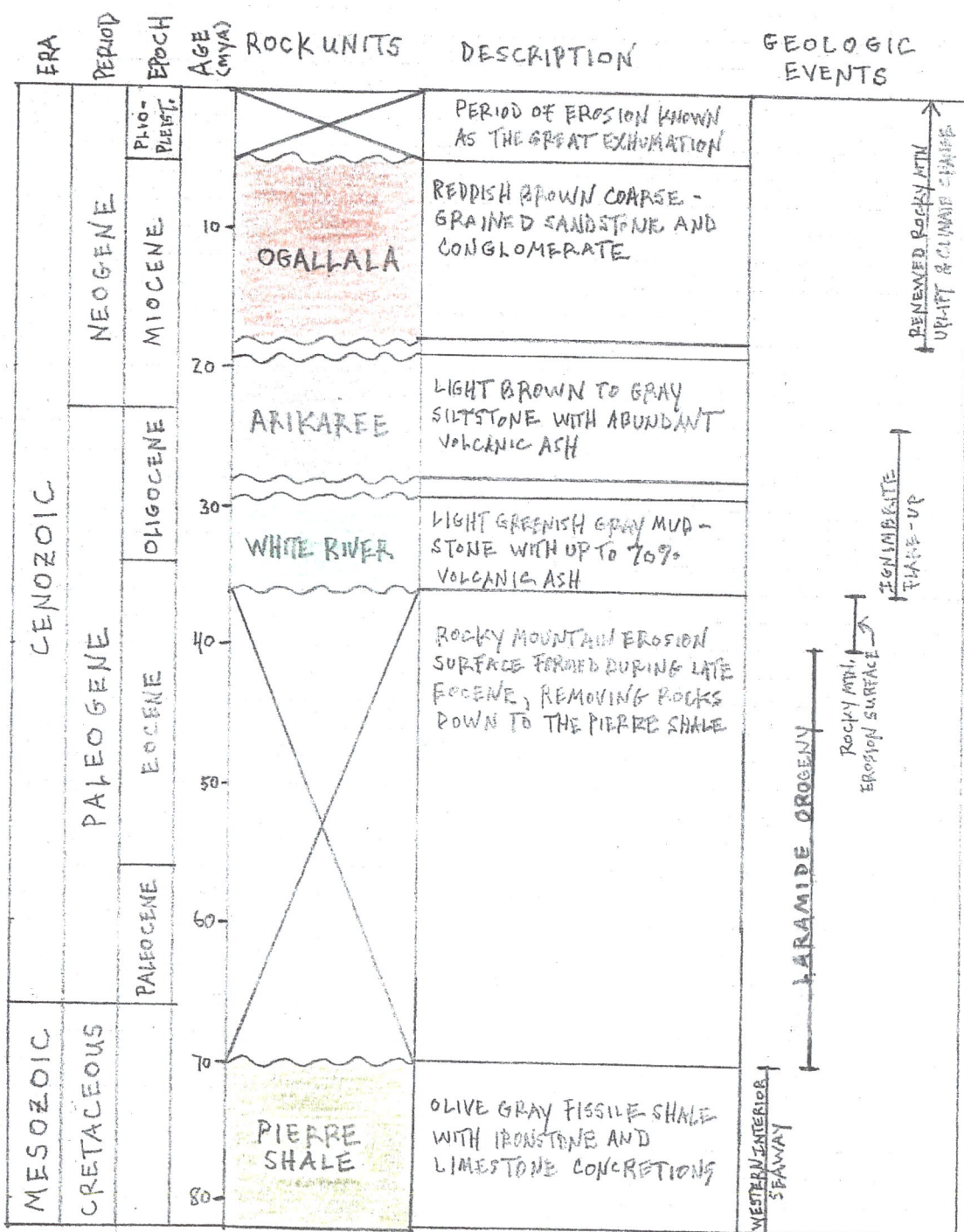

Fig. 75 Stratigraphic column showing age and sequence of rock units at Soapstone Prairie.

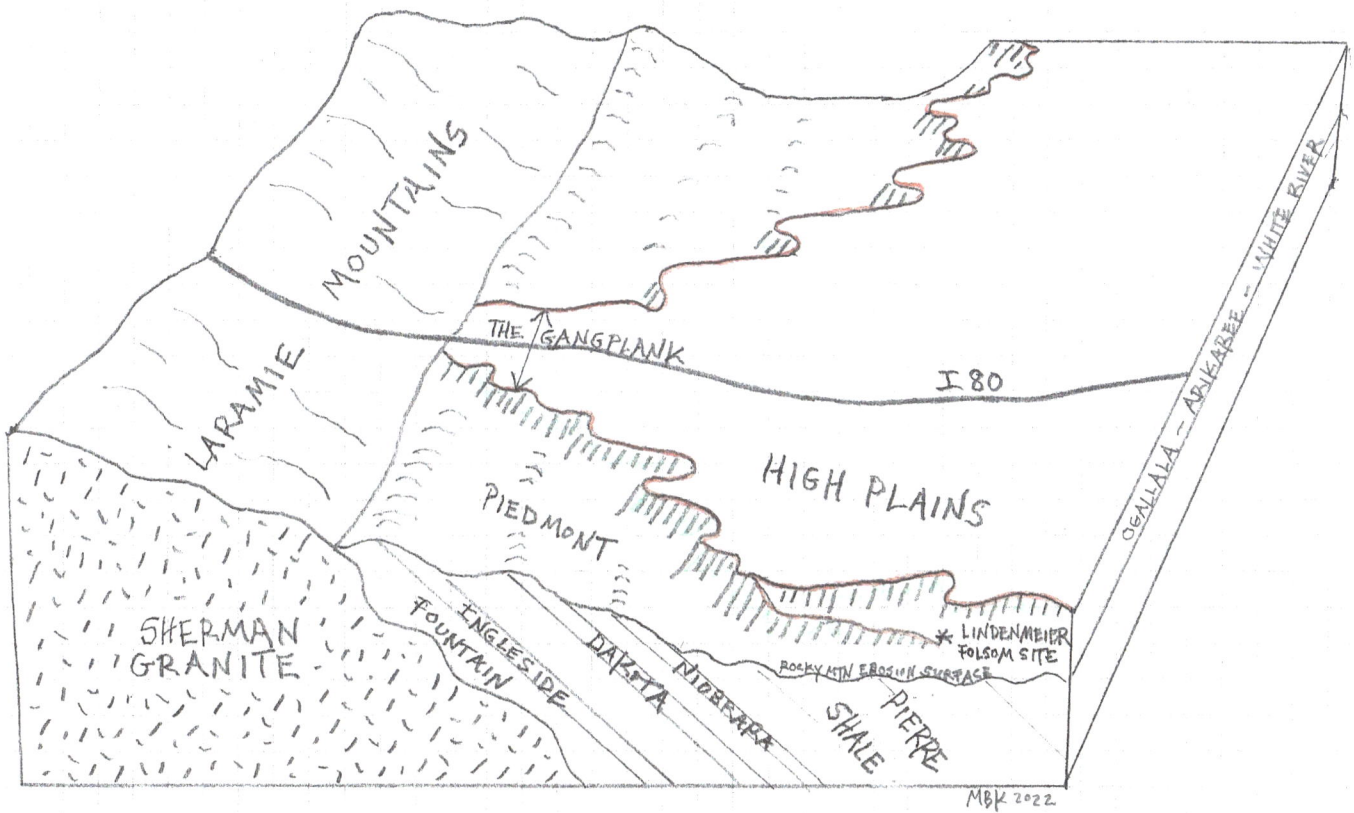

Fig. 76 Block diagram illustrating the geologic setting of the Soapstone Prairie area. The Rocky Mountain erosion surface is Late Eocene in age and cuts across the rock units that were tilted up during the Laramide Orogeny. This surface of contact is known as an angular unconformity because it separates older tilted rocks below from younger flat-laying rocks above.
(Modified from interpretive sign posted on Interstate 80)

Fig. 77 Outcrop of cross-bedded Ogallala conglomerate and coarse sandstone seen off the Mahogany Loop Trail.

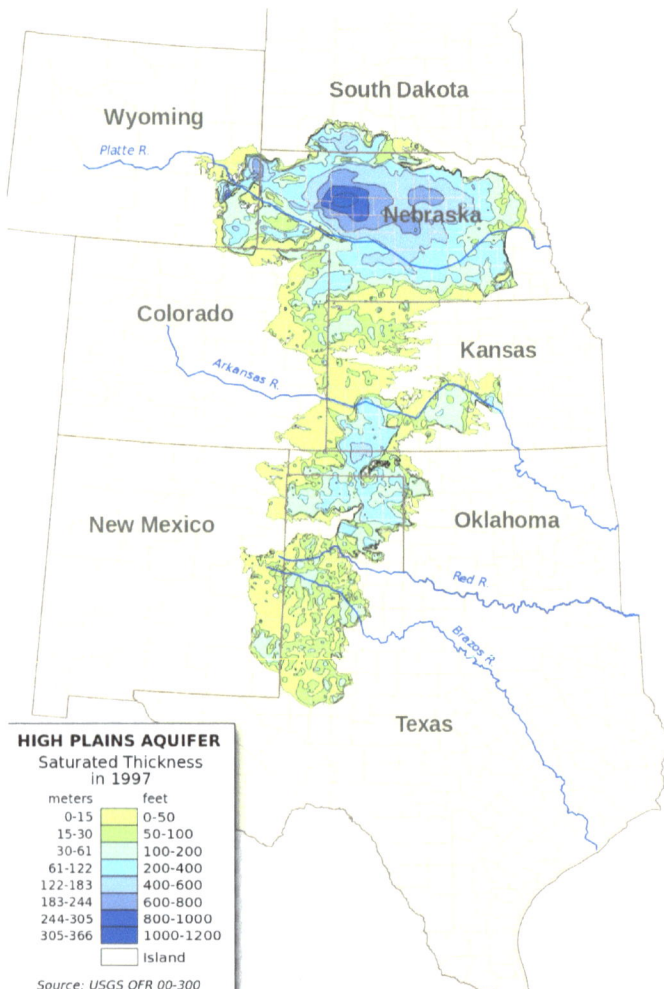

Fig. 78 Geographic extent of the Ogallala aquifer.

HIGH PLAINS AQUIFER
Saturated Thickness
in 1997

meters	feet
0-15	0-50
15-30	50-100
30-61	100-200
61-122	200-400
122-183	400-600
183-244	600-800
244-305	800-1000
305-366	1000-1200
	Island

Source: USGS OFR 00-300

PART D

Geologic Guide

To

Select Natural Area Hikes

Hikers on the K-Lynn Cameron Trail at Red Mountain Open Space. This section presents twelve hikes that cover the full range of rock formations and geologic features discussed in the guide.

HIKE #1
COYOTE RIDGE NATURAL AREA & DEVIL'S BACKBONE OPEN SPACE
Coyote Ridge, Rimrock, Indian Summer Trails
(~ 9 miles & 1,400 feet of ascent—roundtrip to stop 15)

This hike traverses across all the rock units of the northern Colorado Front Range foothills. Beginning in the youngest Pierre Shale, the trail leads west over progressively older rock units, ending within Precambrian metamorphic schist on the east flank of Milner Mountain. Some of the rock units are covered by vegetation or slope debris and are better observed on other hikes highlighted in this section. However the typical topographic expression (ridge, slope, valley) of each of the rock units can be experienced on this hike and can be used as a first order clue on locating yourself geologically on other foothills hikes.

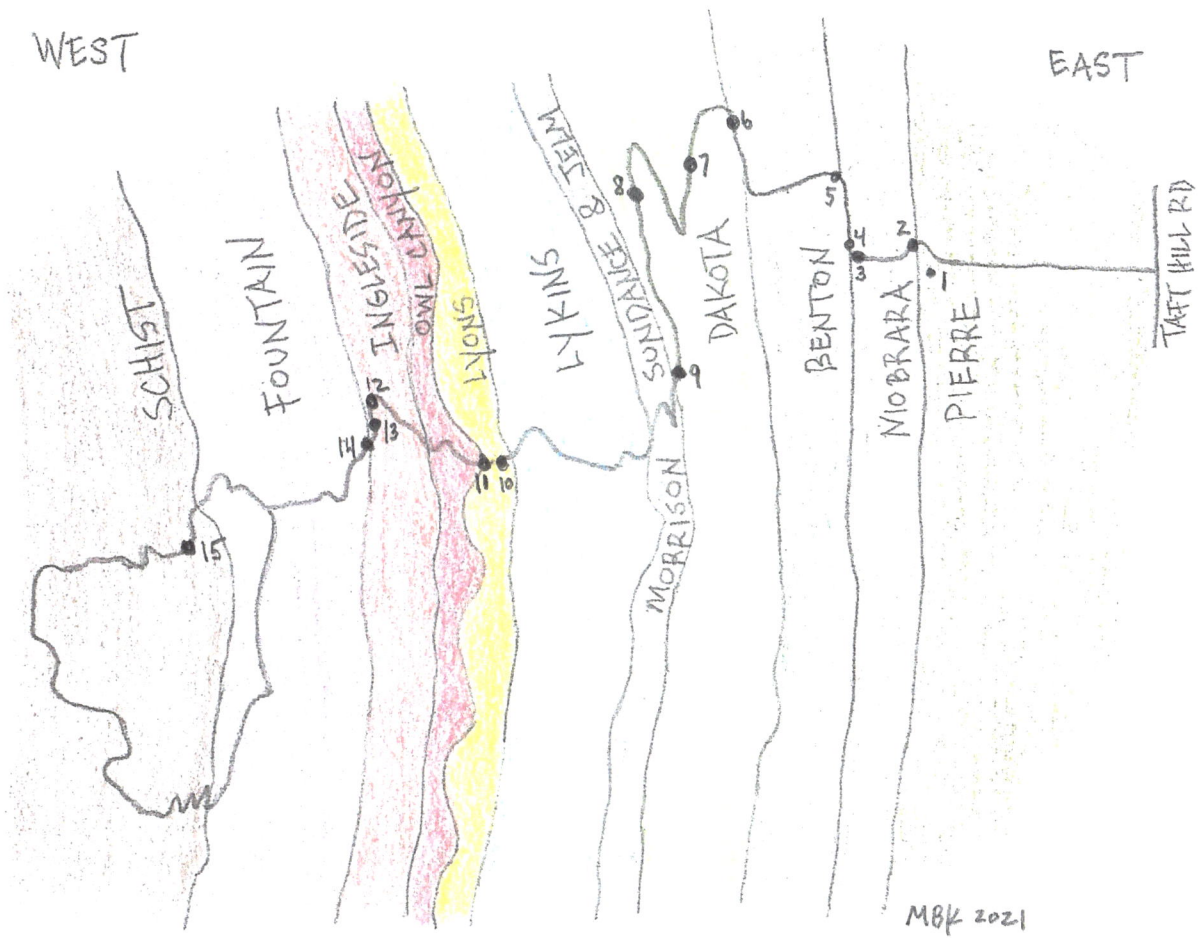

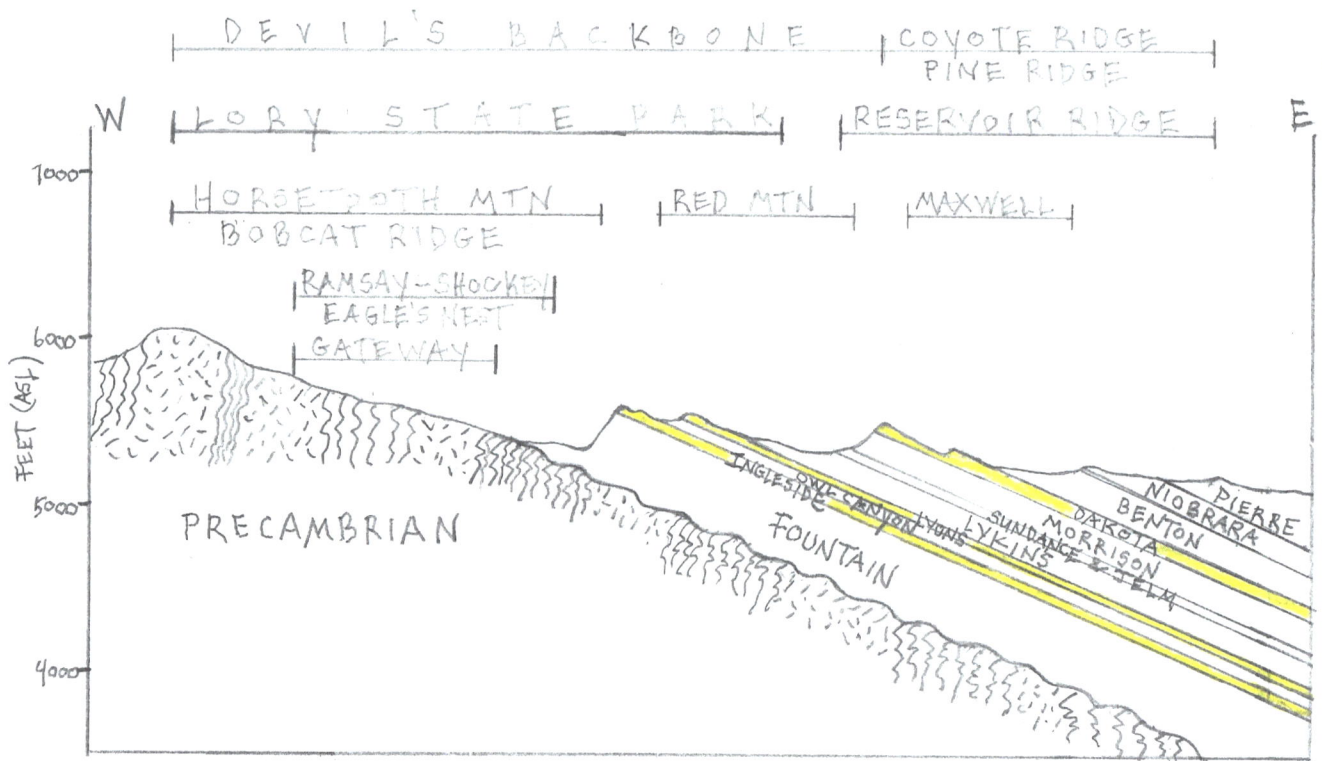

Schematic cross-section illustrating the sequence of rocks this hike traverses starting at the Coyote Ridge trailhead located on top of the Pierre Shale, and ending in Precambrian schist located on the Indian Summer Trail within Devil's Backbone.

NOTES ON TOPOGRAPHIC EXPRESSION, ROCK TYPE AND APPROXIMATE AGE

Pierre Shale: Black flaky shale underlying the western edge of the Great Plains. **72 mya**
Niobrara: Limestone forming low-relief hills and ridges. **82 mya**
Benton: Shale and limestone forming a valley. **89 mya**
Dakota: Sandstone & conglomerate forming two hogback ridges, and shale valley. **100 mya**
Morrison: Mudstone, limestone & sandstone forming west-facing slope. **148 mya**
Sundance & Jelm: Sandstones forming lower slope beneath Morrison. **165 mya & 240 mya**
Lykins: Red mudstone, limestone and gypsum forming a valley. **248 mya**
Lyons: Sandstone forming a hogback ridge. **276 mya**
Owl Canyon: Mudstone, siltstone and sandstone forming a shallow valley. **280 mya**
Ingleside: Sandstone forming a hogback ridge. **285 mya**
Fountain: Conglomerate and sandstone forming a slope and valley. **290 mya**
Basement Rocks: Schist/gneiss/granite. **1700-1800 mya**

1 Flaky black Pierre Shale.

2 Niobrara 'A' chalk.

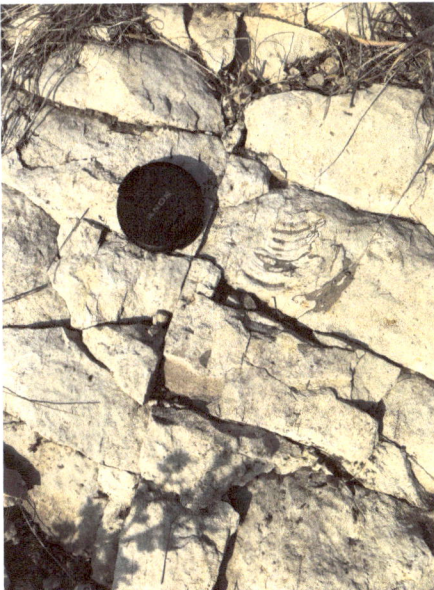

3 Inoceramous fossil in Fort Hayes Limestone member of Niobrara Formation.

4 View looking south from Fort Hayes ridge. At center left is a rare outcropping of Benton Group rocks. The Benton underlays the valley between the Fort Hayes ridge and the Dakota hogback on the far right.

5 Codell Sandstone on slope
beneath Fort Hayes ridge.

6 Upper Dakota (Muddy Sandstone) hogback on right.
Lower Dakota hogback forms rise on left; valley in-
between underlain by Skull Creek Shale.

7 Skull Creek Shale.

8 Ripples and iron concretions in
the Plainview Sandstone.

9 South view from the top of the Lower Dakota hogback ridge at the valley underlain by the Lykins Fm. The vegetated slope below the ridge conceals the Morrison, Sundance and Jelm Formations.

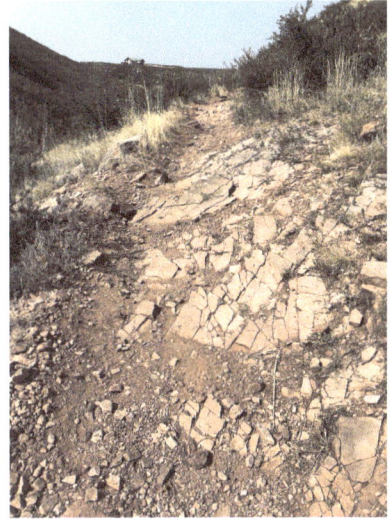

10 Lyons Sandstone out-cropping on the trail.

11 Red beds of Owl Canyon exposed just below contact with Lyons at upper right.

12 Ingleside hogback ridge.

13 View looking south at the Fountain Fm Valley from the top of the Ingleside hogback. Milner Mountain on the right.

14 Contact between the Ingleside and Fountain formations.

15 Outcrop of mica schist on east side of Milner Mountain.

HIKE #2
LORY STATE PARK
Arthur's Rock Trail
(3.2 miles & 1150 feet of ascent—roundtrip)

This popular trail quickly enters Arthur's Rock Gulch from the trailhead bringing you into contact with Colorado's most ancient geologic history. Most of the rocks, both in-place outcrops as well as loose rocks bordering the trail, are circa 1.7 billion-year old igneous rocks known as the Boulder Creek Granodiorite, and a somewhat younger intruded Pegmatite. About two-thirds the way up the trail you will also start to see and traverse over even older metamorphic schist dated at nearly 1.8 billion years old. (For more background geology, read or review the section on Precambrian Crust in Part A of this guide)

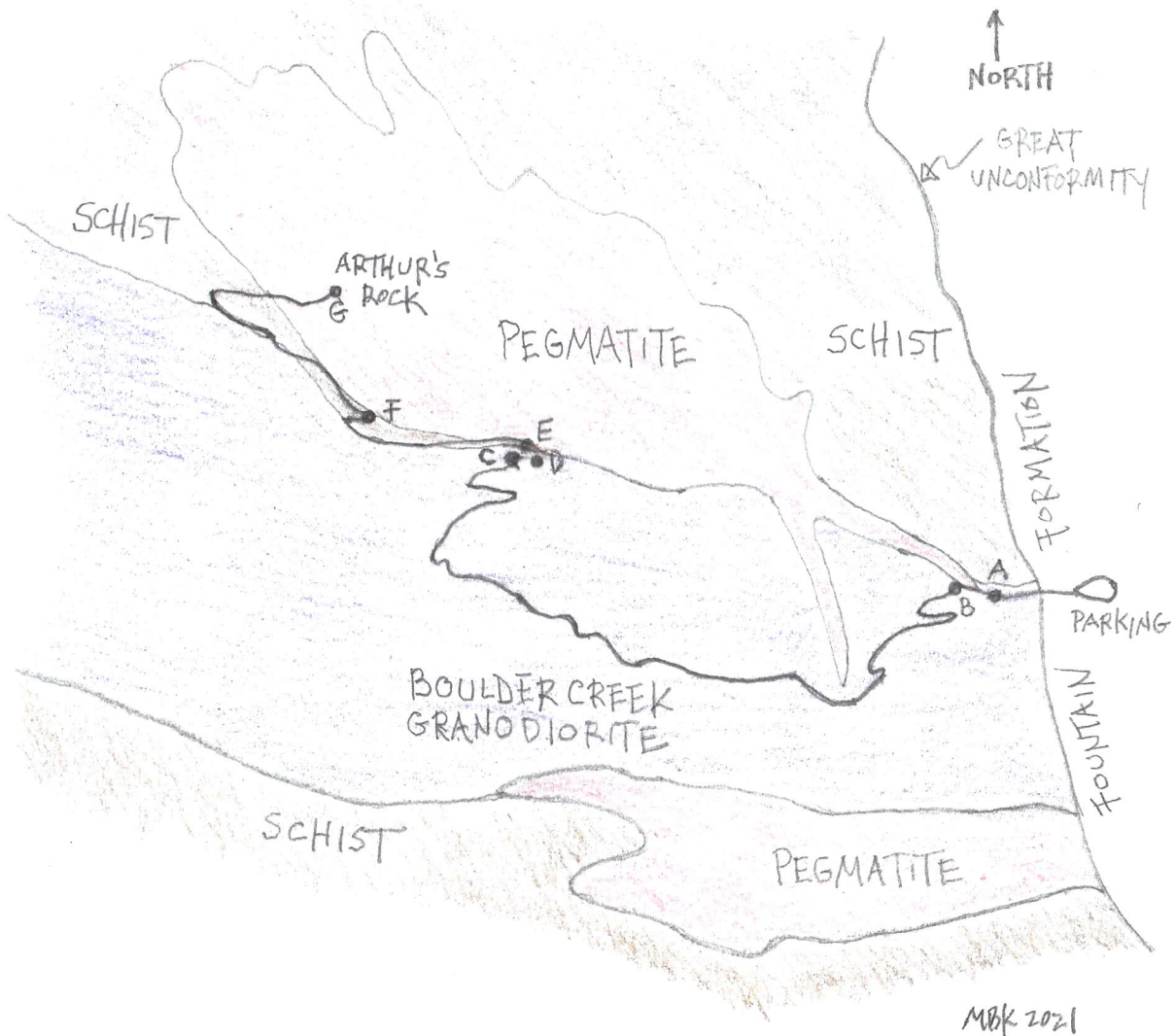

A. Loose boulder of Boulder Creek Granodiorite next to the boardwalk crossing Arthur's Creek. Granodiorite is typicall light to dark gray in color with a salt and pepper appearance.

B. Loose boulder composed of both dark gray granodiorite and pink colored pegmatite. Large potassium feldspar crystals give the pink color.

C. Another look at the contact between the pink colored pegmatite and the gray granodiorite.

D. At overlook point, an outcrop of granodiorite (on the right) and a vein of pink pegmatite cutting through on left.

E. First appearance of metamorphic schist on the trail.

F. Contact between lichen encrusted pegmatite and schist.

G. View of Arthur's Rock pegmatite from the lower trail.

HIKE #3
BOBCAT RIDGE NATURAL AREA
Valley Loop, DR, & Power Line Trails
(8 miles & 1550 feet of ascent—roundtrip)

This figure eight hike begins in the valley underlain by the 300 million year old Fountain Formation. The upper segment of the Valley Loop Trail then follows along the Fountain's western outcrop edge bordering its contact with 1.7—1.8 billion year old metamorphic and igneous rocks. The actual surface of contact, known as the Great Unconformity, is unfortunately concealed by vegetation but adjacent exposures of metamorphic schist and Fountain Fm conglomerates provide guidance on its approximate location. The remainder of the hike traverses across metamorphic schists that are often intruded by thin dikes of granitic rock known as tonalite. At the summit of the hike is the main body of tonalite that stretches across a broad meadow called Mahoney Park. (For more geologic background, refer to sections on the Precambrian Crust, and the Fountain Formation in Part A of this guide)

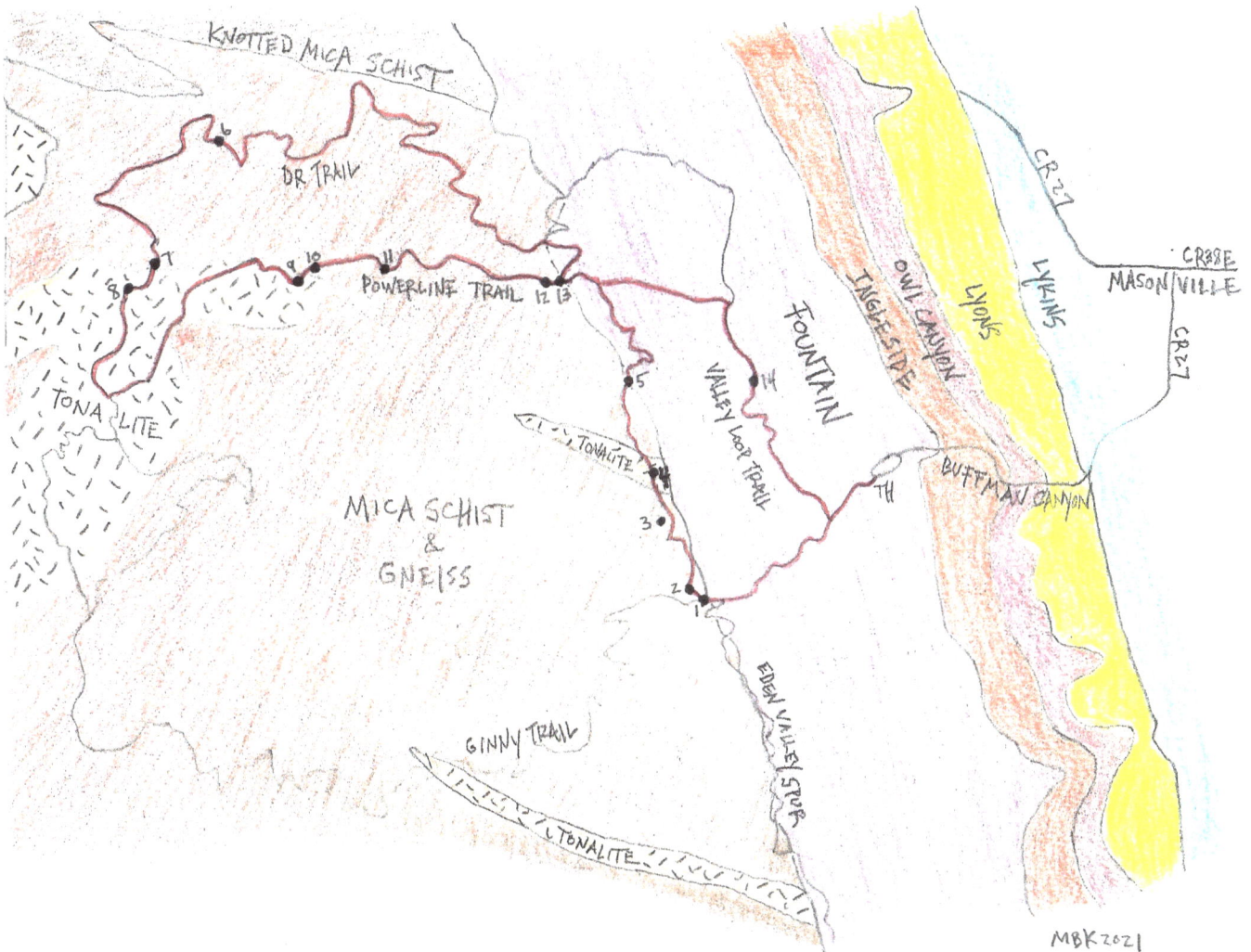

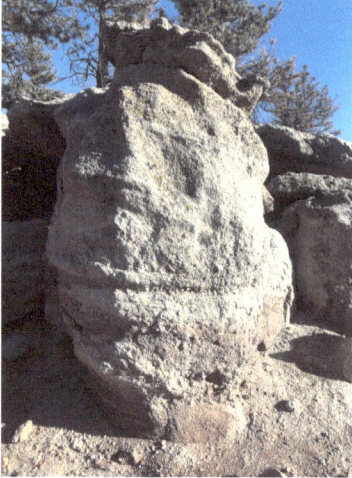

1 Fountain Fm conglomerate and coarse sandstone.

2 Dark mica schist outcrops.

3 Thick milky quartz vein.

4 Tonalite dike weathered into rounded boulders.

5 Easily weathered cobble conglomerate of the
  Fountain Fm.

6 Knotted mica schist with
  large staurolite crystals.

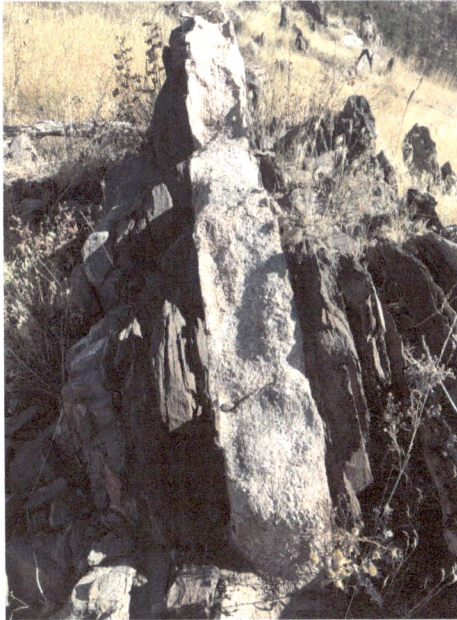

7 Tonalite vein injected into mica schist.

8 Main body of igneous tonalite.

9 Main body of the tonalite intrusion flooring the Powerline Trail.

10 Nice example of three intersecting rock types: dark colored schist cut by purple colored quartz vein, later intruded by light colored tonalite.

11 Large staurolite crystals in a mica schist.

12 Thick bedded sill of tonalite intruded into mica schist.

13 Pseudo Great Unconformity contact of a loose block of Fountain Fm conglomerate Resting on mica schist.

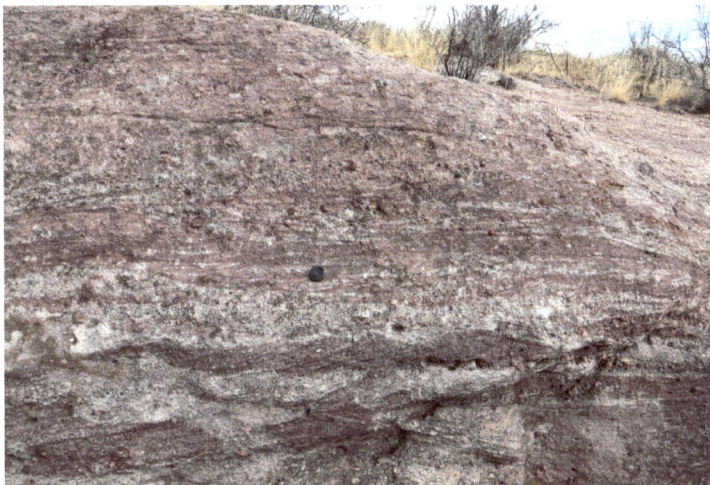

14 Outcrop of cross-bedded pebble conglomerate and coarse sandstone of the Fountain Fm.

Hike #4
LORY STATE PARK
Shoreline & Hogback Overview Trails
(2.5 miles & 300 feet of ascent—roundtrip)

Starting from Arthur's Rock parking lot, this hike focuses on exploring the sedimentary rocks that border the west side of Horsetooth Reservoir. Their typical topographic expression is easy to appreciate here. Well cemented sandstones like the Ingleside and Lyons form hogback ridges, while poorly cemented conglomerate of the Fountain, and easily eroded mudstones of the Owl Canyon and Lykins, form slopes and valleys adjacent to the hogbacks. Although there is some exposure of the Lykins at the shoreline, most of it underlies Horsetooth Reservoir. At the trail terminus it is worth continuing south along the shoreline to see one of the historic Lyons Sandstone quarries. And if the reservoir is drawn down, the headland south of the quarry has an accessible outcropping of stromatolite limestone—part of the Forelle Limestone Member of the Lykins Formation. (For more background geology, read or review sections on the Fountain through Lykins formations in Part A of this guide. Also, see Fig. 3 as a reference cross-section)

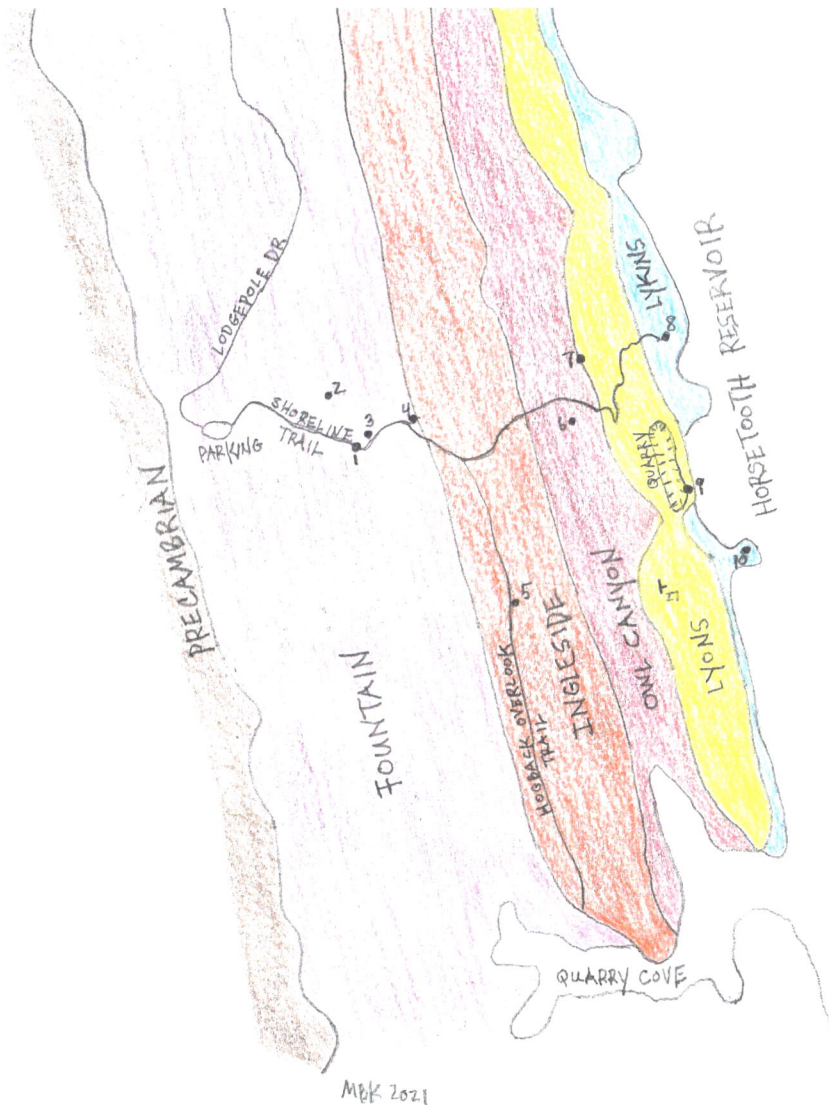

1 View looking south down the valley underlain by the easily erodible Fountain Fm. The Ingleside hogback is on the left and the Precambrian basement is on the right.

2 Large outcrop of Fountain Fm pebble conglomerate composed of cut and fill channels.

2A Outcrop of Fountain Fm exhibiting the change from alluvial fan conglomerate (lower maroon colored layer) to orange colored marine sandstone.

3 Fountain Fm outcrop exhibiting white layers thought to have formed by the flow of hydrocarbon rich waters that reduced the iron mineralization (red coloring).

4 Contact between marine sandstones of the Fountain Fm (rocks extending halfway up the photo with gray splotches) and the overhanging dune sandstone rocks of the Ingleside Fm.

5 Dune cross bedding in Ingleside sandstone.

6 View looking south down the valley underlain by the Owl Canyon Formation between the Lyons hogback on the left and the Ingleside hogback on the right.

6A Wavy bedding in a thin outcropping of Owl Canyon sandstone.

7 Dune cross bedding in an outcrop of Lyons Sandstone along the hogback ridge.

8 Lykins Formation red mudstone outcropping at shoreline.

9 Historic Lyons Sandstone quarry. Dune cross-bedding is well displayed.

10 Outcropping of stromatolitic Forelle Limestone on an emerged headland when Horsetooth Reservoir is drawn down.

RED MOUNTAIN OPEN SPACE
Bent Rock & K-Lynn Cameron Trails
(6.5 miles & 700 feet of ascent—roundtrip)

At more than 15,000 acres, remote Red Mountain Open Space (RMOS) is a scenic and geologic wonderland. RMOS sits at the northern end of an area referred to as the Big Hole, a broad valley highlighted with highstanding buttes and hills where the rocks are folded into anticlines and synclines. The trails featured on this hike provide a walking tour past many interesting rocks, fossils, and prominent landscape forms. For one, the red rocks are of the Owl Canyon and Lykins formations which are usually concealed by vegetation within the other natural areas. Of added interest are gypsum and alabaster rocks at a historic quarry, fossil stromatolite outcrops, views of inverted topographic highs formed by synclinal folds, the opportunity to walk through the Sand Creek Anticline (locally known as Bent Rock), witnessing that Sand Creek cuts through the anticline rather than course around it, and walk past several Ice-Age gravel terraces within the Sand Creek drainage. (For more background geology, read or review the sections on the Lykins through the Ingleside formations. Also, refer to Fig. 29 for a reference geologic cross-section)

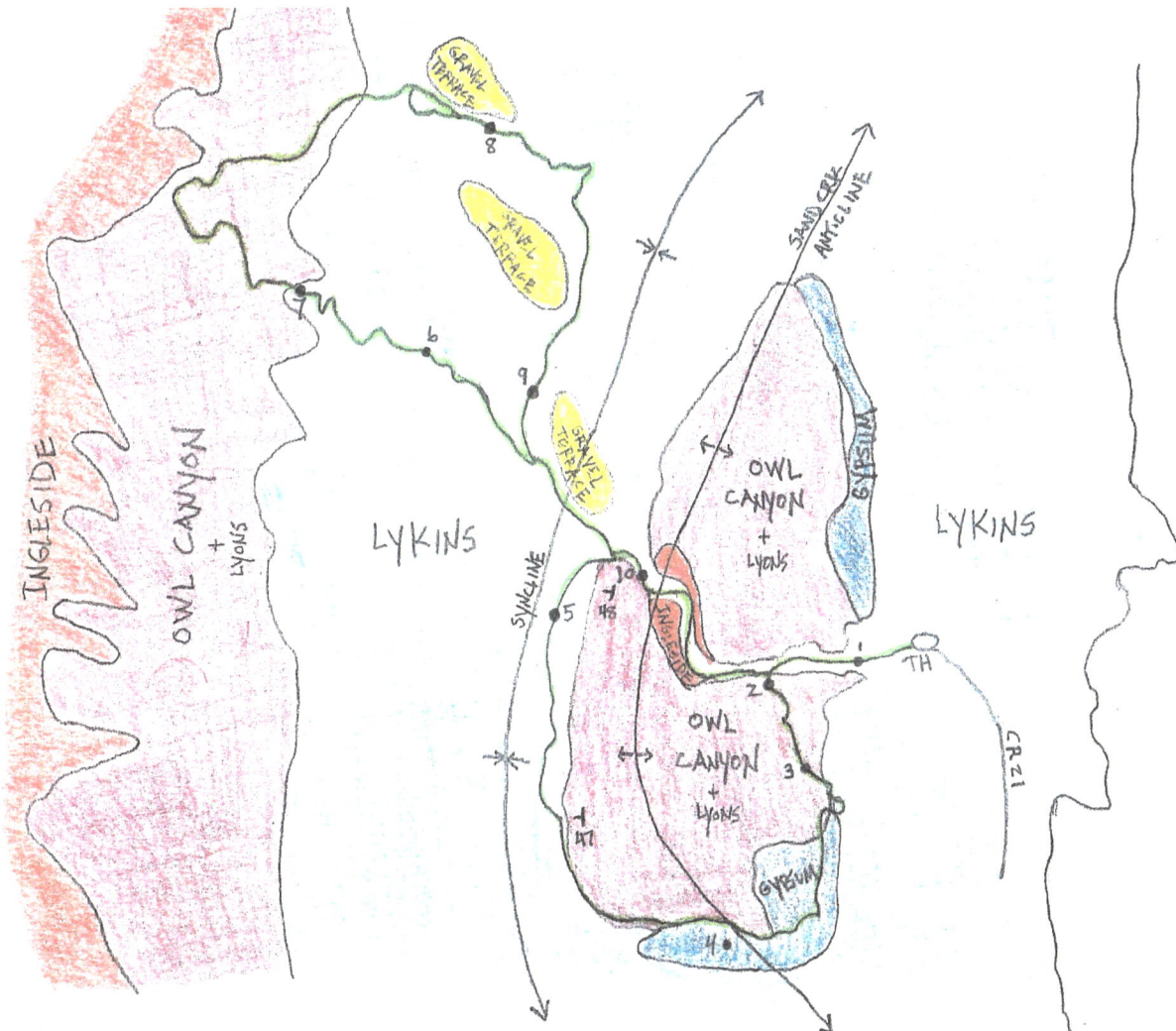

1 View looking west from near the trailhead at the canyon cut by Sand Creek into the Bent Rock anticline.

2 Boulders of granite deposited on the Sand Creek floodplain when base level was much higher and near the top of the anticline.

3 Outcrop of cross-bedded Lyons Sandstone overlying contact with Owl Canyon Fm.

4 Blocks of anhydrite at the site of an old quarry. Gypsum/anhydrite beds are found near the base of the Lykins Fm.

5 View looking north at an exposed synclinal fold, an example of inverted topography.

6 Cross-sectional view of a finely laminated stromatolite in the Forelle Limestone member of the Lykins Fm.

7 Calcite mineralization in a vuggy, brecciated zone created by the leaching of gypsum beds within the Lykins Fm.

8 Top view of circular to oval shaped stromatolite mounds within the Forelle Limestone.

9 Gravel terrace deposited by Sand Creek on an Ice-Age floodplain.

10 View looking into Sand Creek Canyon from the west. The deep-red, thinly bedded Owl Canyon Formation on the right rests on the thickly bedded limestones and sandstones of the Ingleside Formation.

HIKE #6
DEVIL'S BACKBONE OPEN SPACE
Morrison, Wild & Hunter Loop Trails
(4.4 miles & 600 feet of ascent—roundtrip)

First off it is worthwhile taking the short 0.8 mile Morrison Trail Loop south of the parking lot for its well-done geologic displays and interpretive signs. Here you will learn (among other things) that Devil's Backbone is a product of geologic structure modified by weathering and erosion. The structure is known as the Milner Mountain Anticline, a basement cored uplift that resulted in folding of the overlying sedimentary rocks into an anticline during the Laramide Orogeny. The anticline is asymmetric, that is, the rocks on the west limb incline very steeply, dipping around 80 degrees to the west while the rocks on the east limb are inclined gently to the east at around 15 degrees. This trail explores the steeply dipping west flank of the anticline where differential erosion has left a standing wall of Dakota rocks forming the 'backbone'. The hike proceeds to the core of the anticline with exposures of older sedimentary rocks and a view of the changing dips on opposite limbs of the anticline. (Refer to Fig. 57b for a geologic cross-section through the anticline)

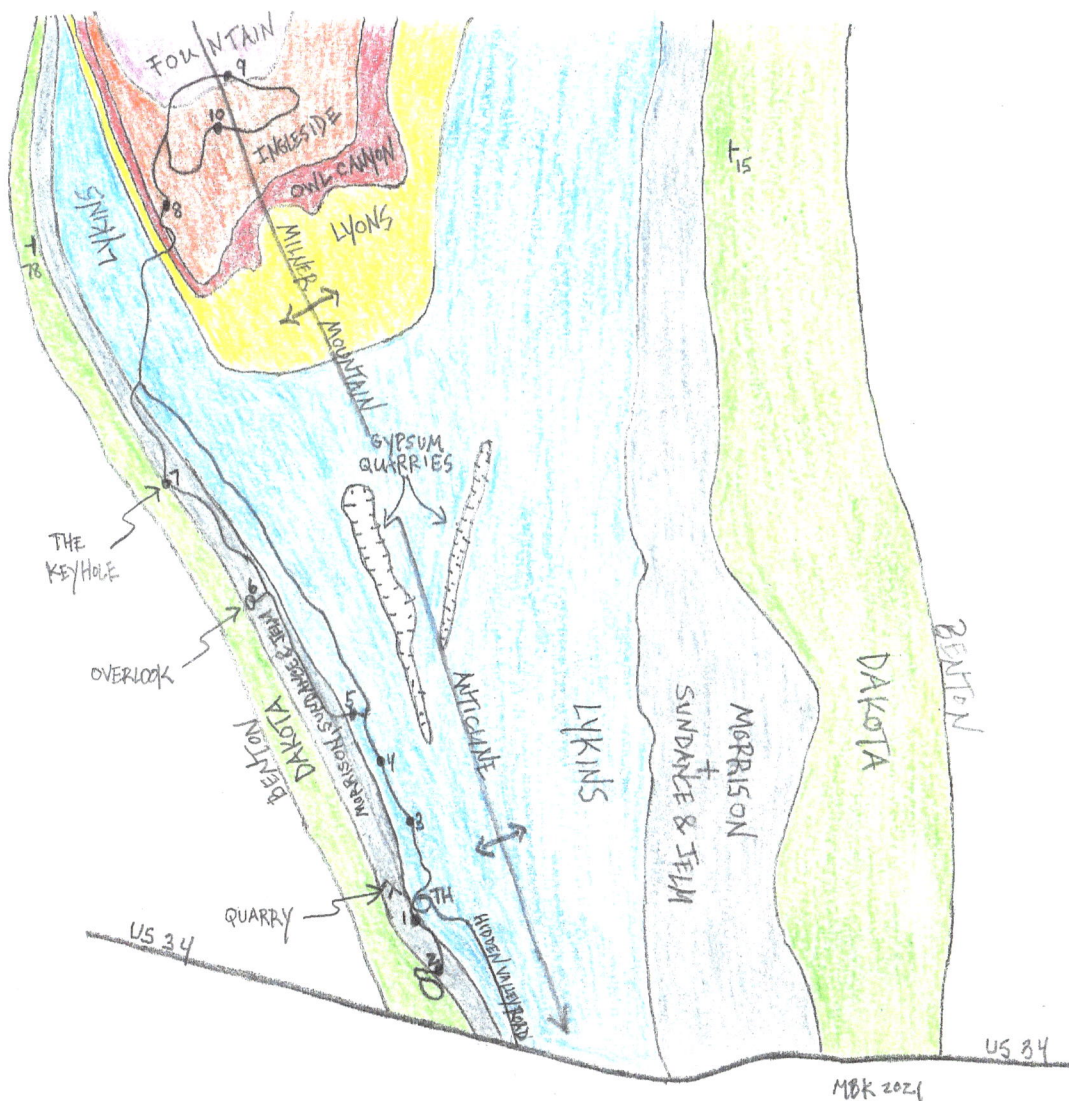

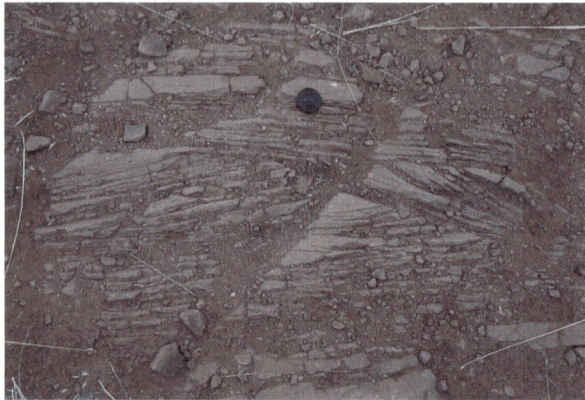

1 Cross bedded Sundance Formation on the Morrison Trail. The Sundance is an ancient sand dune deposit.

2 View looking north from the Morrison Trail. Steeply dipping beds of the Dakota form the Devil's Backbone. Historic Morrison quarry on the right.

3 Trailside exposure of steeply inclined west-dipping beds of the Lykins Formation.

4 View of the historic gypsum quarries in the center of the photo.

5 Cross bedded Sundance Formation on Wild Loop Trail.

6 Gray limestone beds of the Morrison Formation outcropping at the Overlook.

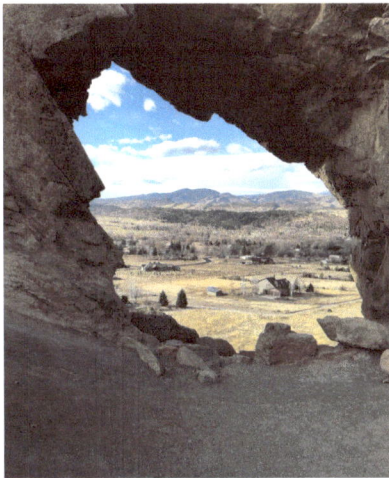

7a The Keyhole—a window eroded through the Devil's Backbone.

7b Close-up of the resistant conglomerate that forms the Dakota Backbone. Geologists refer to this part of the Dakota as the Lytle Formation.

7c View of the contact between the Dakota on the left and the colorful Morrison Formation on the right.

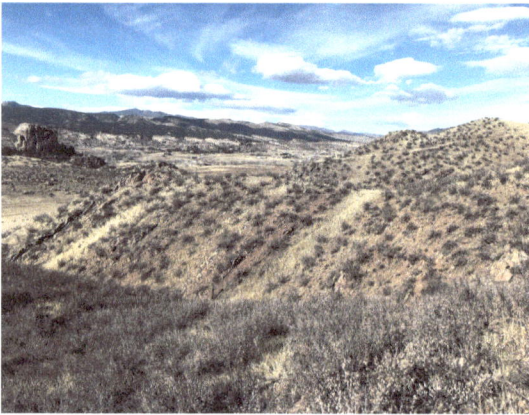

8 View looking northwest at the west dipping beds in the middle of the photo of the Lyons, and Owl Canyon formations.

9 View of the contact between the Fountain and Ingleside Formations. The Fountain is slightly recessed, dull maroon colored, and overlain by the orange colored Ingleside.

10 Calcite filled fractures in the Ingleside Formation near the axis of the anticlinal fold.

Hike #7
MAXWELL NATURAL AREA
Foothills Trail to Aggie Peak
(4 miles & 630 feet of ascent—roundtrip)

Rockslides associated with the Dakota hogback have been dated as occurring circa 30 million years ago to as recently as several hundred thousand years ago. On this hike you will walk across the boulder-strewn hummocky topography of a rockslide debris fan, observe the east facing escarpment created by the detachment thrust fault, and observe how the hogback ridgeline has been displaced leaving a widened valley in its place..The Aggie Peak segment of the trail takes you to the top of the undisturbed part of the hogback where you can examine rocks of the Muddy Sandstone, the uppermost member of the Dakota Group. The Muddy Sandstone is an important oil and gas producing reservoir within the Denver Basin, and famous for its dinosaur tracks at Dinosaur Ridge in Morrison, Colorado. (See Figure 56 in the guidebook for a schematic model of a Dakota rockslide)

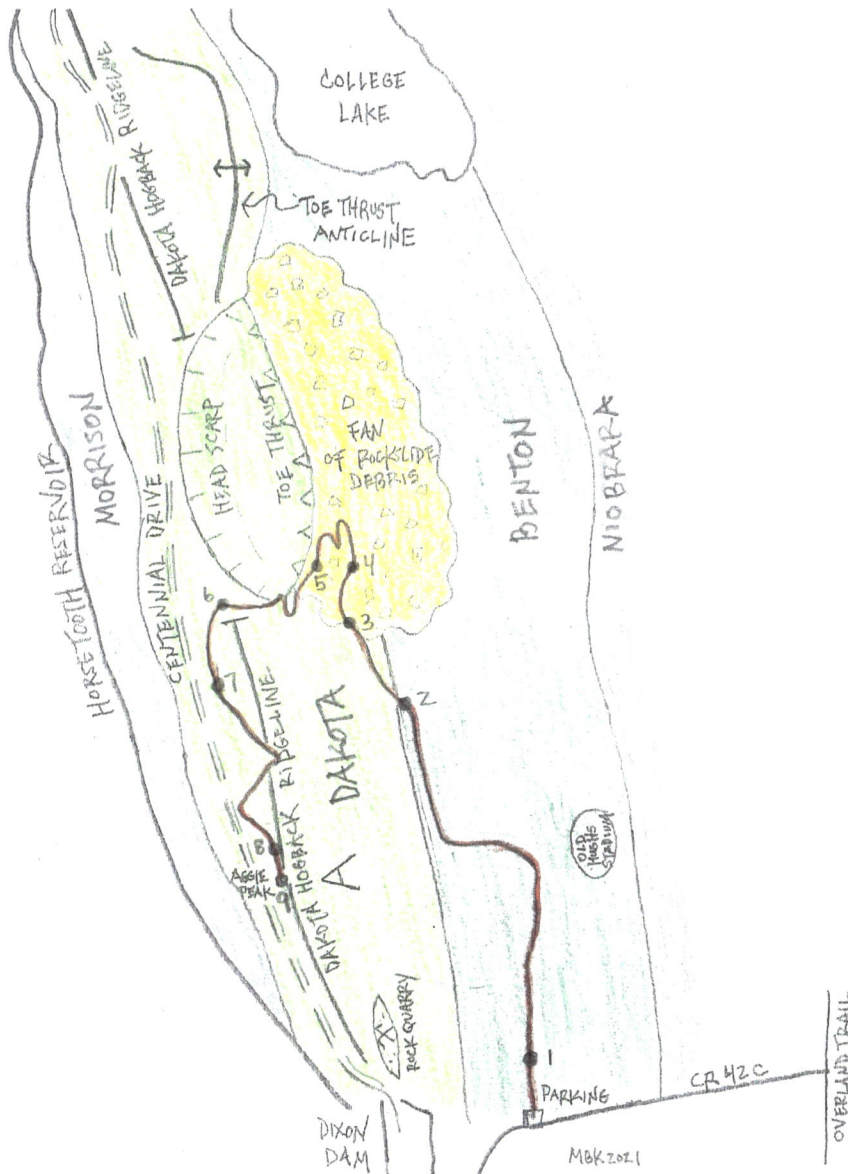

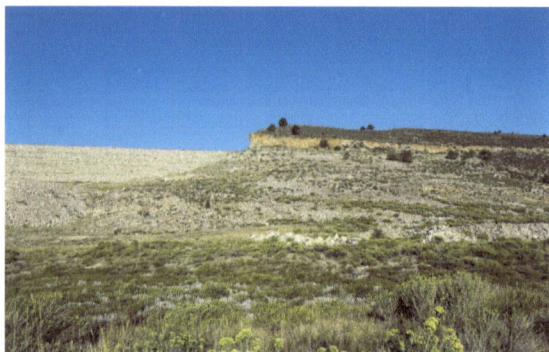

1 View looking west toward the Dakota hogback from near the trailhead. Below the top of the ridge—to the right of Dixon Dam—a quarry was cut into the dip slope of the hogback, exposing the Muddy Sandstone member of the Dakota.

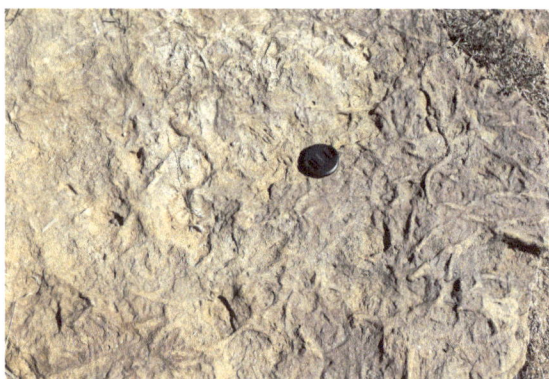

2 This big boulder of Muddy Sandstone is full of trace fossils reflecting the burrowing activity of a shrimp-like crustacean.

3 View looking south from the trail towards Pine Ridge NA. The sinuous ridge is the Fort Hayes Limestone of the Niobrara Fm.

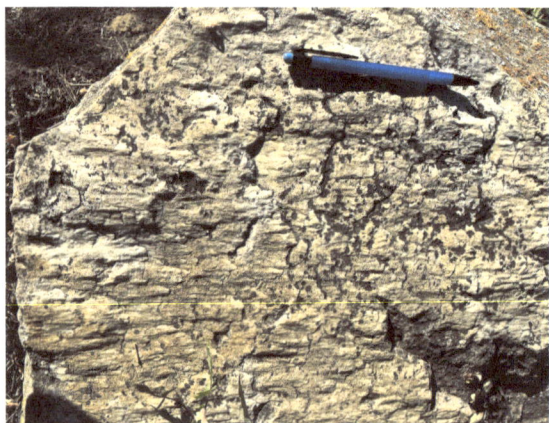

4 An example of **slickensides**, common in the Dakota, which are bedding plane lineations caused by the frictional movement of rock sliding against rock.

5 View looking up towards the east-facing ridge of fractured but intact Dakota sandstone rocks which slid downslope and formed an apron of debris in front of the detachment fault located beneath the ridge.

6 View looking north at the wide meadow formed by the breach of the Dakota hogback ridgeline due to the rockslide detachment.

7 Trailside outcrop of the Plainview Sandstone Member of the Dakota Group. The Plainview Sandstone is thought to be the first rock unit deposited within the Western Interior Seaway.

8 This loose Muddy Sandstone boulder appears to have an imprint similar in shape to a tridactyl dinosaur footprint. Not too far-fetched given that the famous dinosaur trackway at Dinosaur Ridge is in the same Muddy Sandstone rocks.

9 View looking south from the top of Aggie Peak. Centennial Road on the left can be seen tracking through the Skull Creek Shale valley, separating the Muddy Sandstone hogback on the left from the Plainview Sandstone hogback on the right. Underlaying the Plainview, and forming part of the cliff face, is the Lytle Formation. This entire sequence of rock units makes up the Dakota Group.

HIKE #8
RESERVOIR RIDGE NATURAL AREA
Michaud Spur to North & South Loop Trails
(4.8 miles & 650 feet of ascent—roundtrip)

At the Michaud Lane trailhead you will be standing on exposures of gray limestone from the Fort Hayes Member of the Niobrara Formation. The Fort Hayes forms a low-relief ridge that wraps around the east bank of Claymore Lake—which rests on rocks of the valley-forming Benton Group. Also overlying the Benton are debris fans from at least three separate rock slides that detached off the backside of the Dakota hogback. Once above the rockslide landscape, the hike offers great views of the upper and lower Dakota hogback ridges with the intervening valley underlain by the Skull Creek Shale. Significantly, this is the only natural area that affords the opportunity to walk across, and along a fault trace that cuts through the sedimentary rocks. The Bellvue Fault is a thrust fault whereby rocks on the east side of the fault have moved up and slightly over the rocks on the west side of the fault (see cross-section at the end of this section). Interesting landscape changes expressed by the fault include a repeating of the lower Dakota ridge above and below the trail between stops 7 and 8, and an unusual flat laying meadow developed on the backside of the upper Dakota hogback in the vicinity of stop 11.

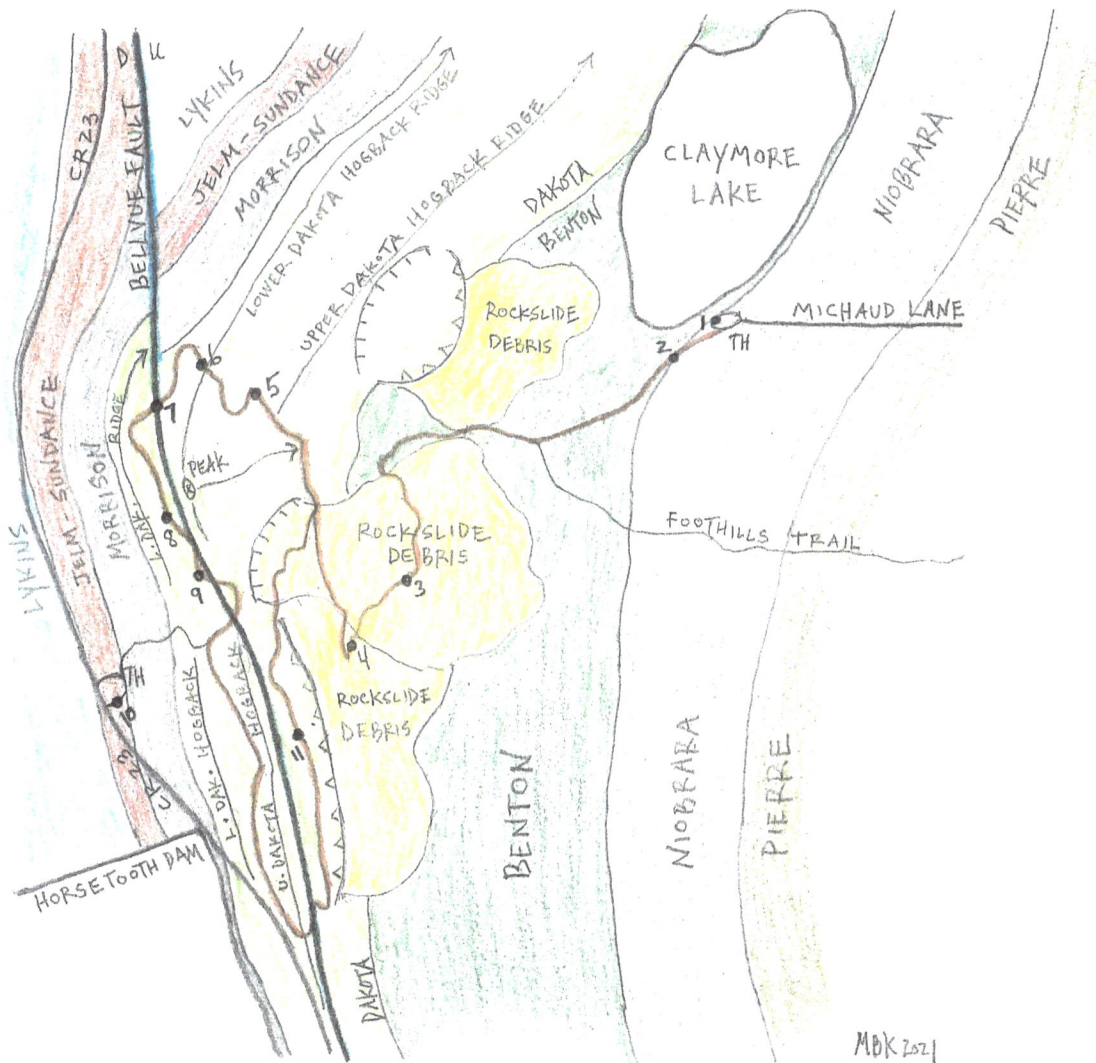

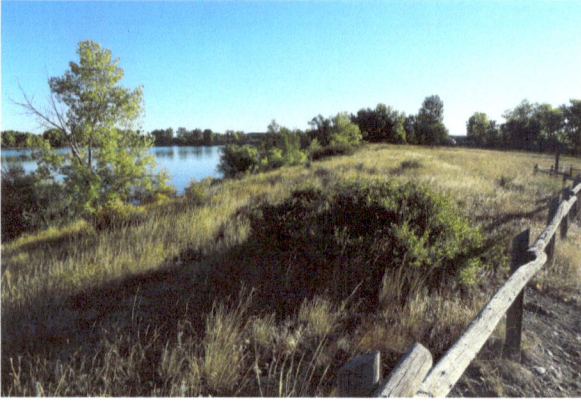

1 At the Michaud trailhead; Claymore Lake on the left is situated within the valley underlain by the Benton Group rocks. The east bank of the lake is formed by the low ridge of the Fort Hayes Limestone—a member of the Niobrara Fm.

2 A look at exposed rock of the Fort Hayes Limestone.

3 The fan of rockslide debris forms hummocky topography.

4 A look at the east facing escarpment formed by a rockslide on the south end of Reservoir Ridge.

5 Looking northeast down the Skull Creek Shale valley sandwiched between the Upper Dakota hogback on the right (Muddy Ss), and just off the picture to the left, the lower Dakota hogback (Plainview Ss and Lytle Fm.).

6 Planar cross bedding in the Plainview Sandstone exposed on the Lower Dakota hogback ridge.

7 View looking along the Bellvue Fault trace toward Bellvue Dome.

8 View looking south from the north end of Horsetooth Reservoir. The tilted beds on the left are hogbacks formed by Dakota Group sandstones while the hogbacks on the right are formed by resistant sandstones of the Lyons and Ingleside formations.The reservoir sits on the valley-forming Lykins Formation.

9 Looking west toward the Precambrian mountains, the valley below is floored by Lykins Fm bedrock. Towards the middle of the valley is the Charles Hansen Feeder Canal that flows to the Cache La Poudre River.

10 Looking up from the west parking lot trailhead off Centennial Drive; the trace of the Bellvue Fault, a west verging thrust fault, marks the fault line where Lower Dakota rocks were thrust upward. As a result, the Lower Dakota forms two ridges as annotated in the photo. (Up arrow marks the highside of the fault)

11 An interesting valley has formed on the backside of the Upper Dakota hogback along the trace of the Bellvue Fault.

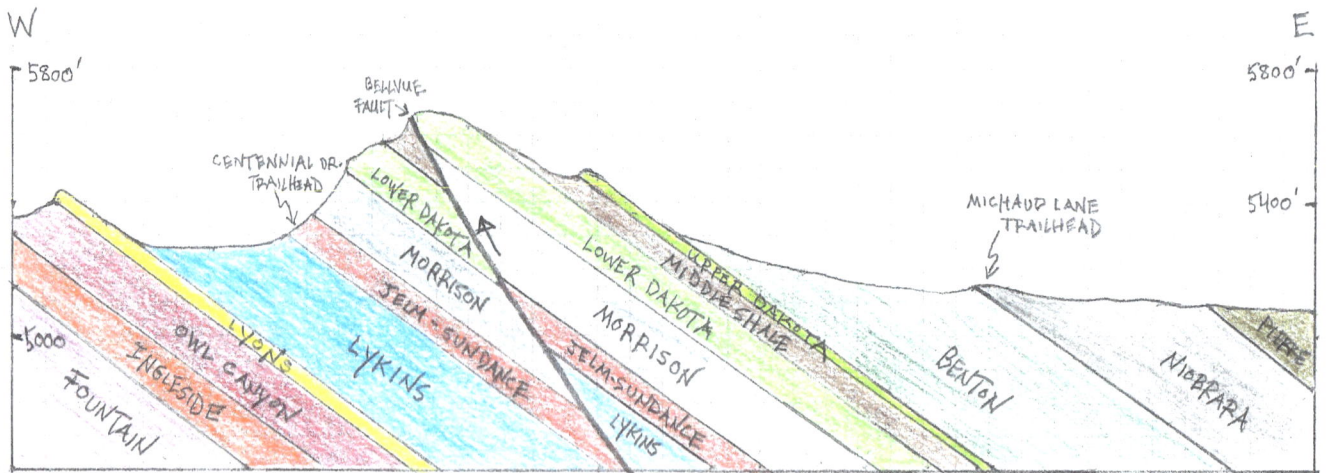

HIKE #9
SOAPSTONE PRAIRIE NATURAL AREA
Lindenmeier Overlook & Mahogany Loop Trails
(3.8 miles & 450 feet of ascent—roundtrip to stop 6)

Soapstone Prairie is Fort Collins biggest prize in its natural areas collection. It is noted for sheer size (29 square miles), expansive views, its reintroduced bison and black-footed ferrets, and the archaeologically important Lindenmeier Folsom Site. Trails leading from the north parking lot are the best way to explore Soapstone's unique geology. From here you will have a chance to examine the light colored, volcanic ash-rich rocks of the Arikaree and Brule formations as well as the Ogallala caprock that form the high plains plateau locally known as the Cheyenne Ridge, or what geologists frequently call 'the gangplank'. (For more geologic background refer to the section on Soapstone Prairie in Part C of this guide)

1 View looking southwest from the Lindenmeier Overlook to the gulch where Paleo-Indian artifacts were discovered and later excavated by the Smithsonian Institute.

2 Just off the Lindenmeier Trail is a hillside exposure of the volcanic ash-rich Arikaree Formation. Along with the underlaying Brule Formation, these light colored rocks are exposed on the south slope of the Cheyenne Ridge and on a clear day can be seen from Fort Collins 25 miles away.

3 Trailside exposure of the Arikaree, and further downslope, the Brule formations. Both these rock units were deposited by streams carrying a heavy load of volcanic ash derived from super volcanoes that were erupting from as far away as Nevada. The rocks from both formations are chalky in appearance and difficult to tell apart. The Brule has a slight greenish-gray cast while the Arikaree is very light brown to white in color.

4 Interesting blocks of limestone are seen here resting on the Ogallala and Arikaree formations. The blocks are erosional remnants from a limestone unit that may have originally formed a caprock overlaying the Ogallala.

5 A close-up look at one of the limestone blocks reveals cobbles and boulders of granite encrusted with a cementing limestone. Other limestone blocks exhibit laminations, sometimes with a mounded appearance. The limestones remain a mystery with respect to their depositional environment, and whether they are of biogenic or non-biogenic origin.

6 Trailside exposure of the Ogallala Formation. As seen here, the sandstones and conglomerates making up the Ogallala are easily eroded due to poor cementation. The Ogallala's thickness, porosity and permeability make it an important and excellent aquifer in the subsurface.

HIKE #10
HORSETOOTH MOUNTAIN OPEN SPACE
Swan Johnson-Shoreline-Sawmill-Loggers-Towers-Herrington Trails
(7.2 miles and 1350 feet ascent—roundtrip)

Starting at the Soderberg trailhead this hike will involve connecting with segments of several different trails that form a somewhat complex network through Horsetooth Mountain Open Space. It may be helpful to have a gps trail tracking app to make all the right connections. The route is designed to see all the different rock types present at Horsetooth. The first two segments are situated within the valley underlain by the Fountain Formation. The Fountain is the oldest sedimentary rock in the Front Range foothills and rests on various Precambrian-aged igneous and metamorphic rocks. Ascending out of the valley and onto the Sawmill Trail you will begin to see outcrops of igneous trondhjemite, pegmatite, and metamorphic schist. Because of the scale at which the U.S. Geological Survey maps, capturing every single outcrop occurrence of each rock type is not feasible. The map below is therefore generalized. And so at stops 2 and 3, which are mapped within the trondhjemite area, there are also occurrences of pegmatite and schist. In fact, the most conspicuous outcrops across Horsetooth Mountain Open Space are pegmatites. Stop 7, the last for this hike, is a good lunch stop where you have good views across the Spring Creek valley towards Horsetooth Mountain. Because of its popularity Horsetooth Rock Trail is also shown on the map in purple.

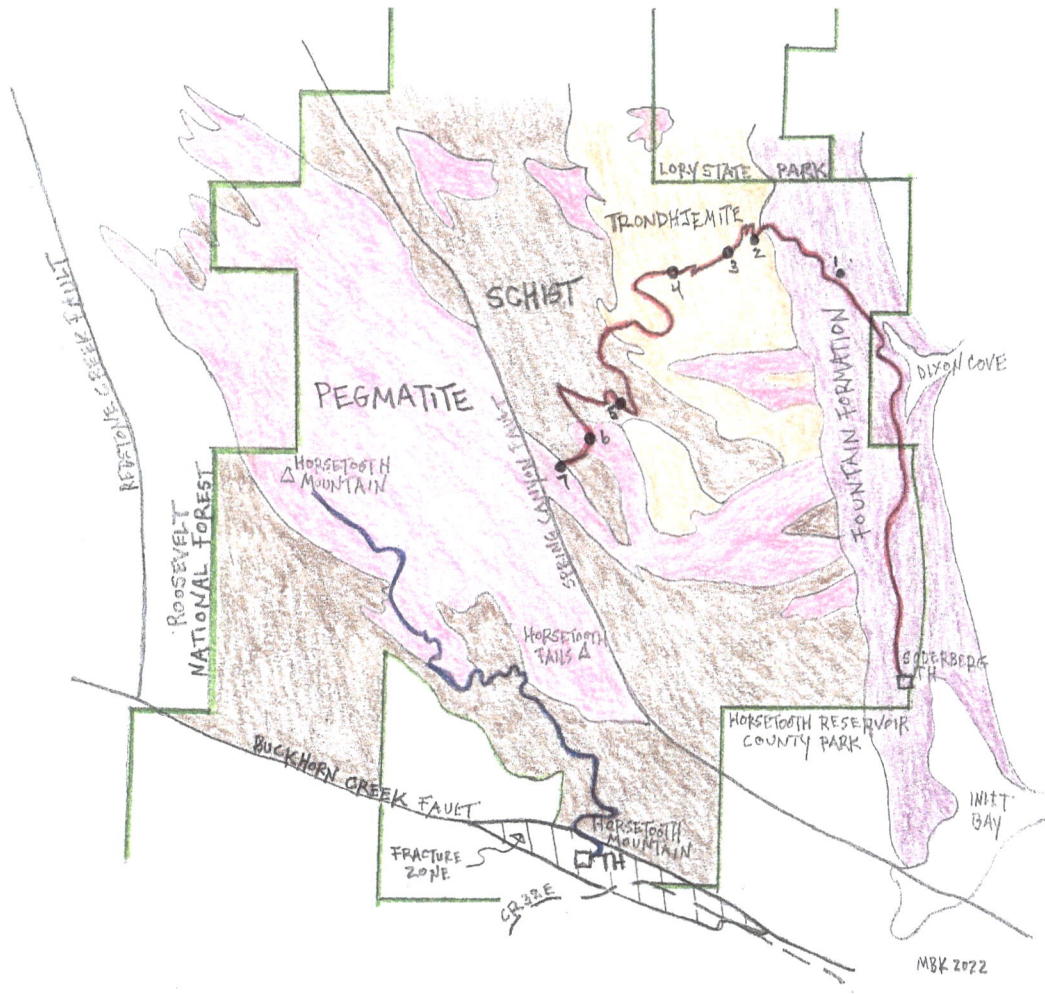

1 Cross bedded Fountain Formation.

2 Outcrop showing contact between fine grained trondhjemite with coarse grained pegmatite.

3 Outcrop of pegmatite in contact with schist.

4 Outcrop of trondhjemite.

5 Weathered schist on eroded hill slope.

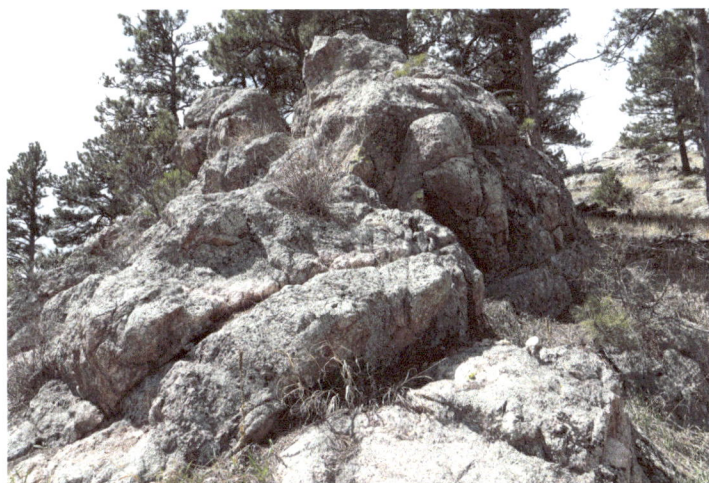

6 Outcrop of coarse-grained pegmatite.

7 West view across Spring Creek Valley towards Horsetooth Mountain. The rock exposures are all pegmatite.

HIKE #11
RAMSAY-SHOCKEY OPEN SPACE
Fisherman's Cove, Besant Point & Shoshone Trails
(6 miles & 300 feet of ascent—roundtrip)

The southernmost of the natural areas within Larimer County, Ramsay-Shockey offers another look at the crystalline basement rocks of the Front Range foothills. But the highlight is that along the banks of a headland on the west side of Pinewood Reservoir you can actually put your finger on the Great Unconformity. Here the 300 million year old Fountain Formation rests directly on top of 1.8 billion year old schist. Also well exposed is 1.7 billion year old igneous rock known as trondhjemite. (For more geologic background, refer to the section on Precambrian Crust in Part A of this guide)

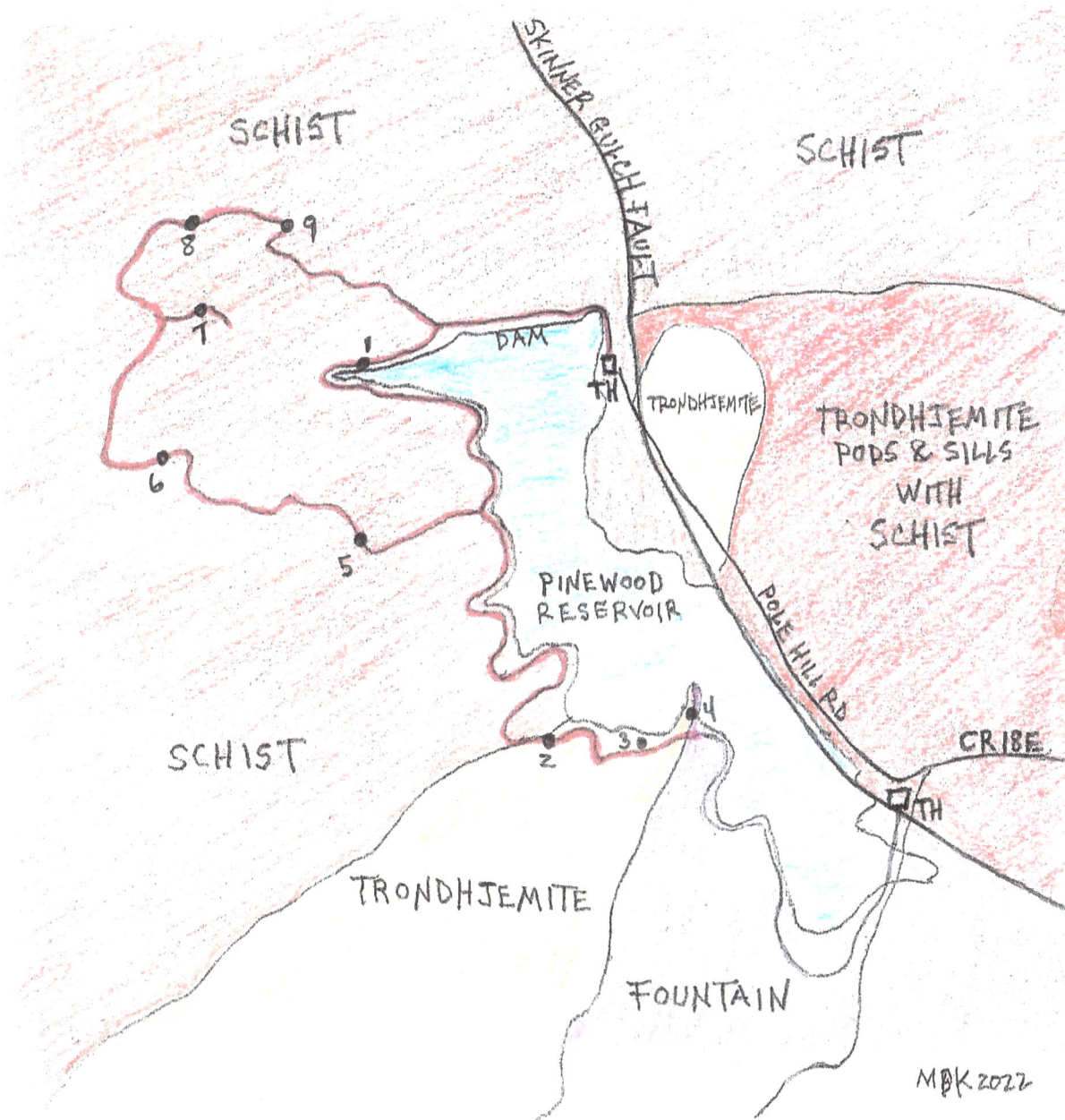

1 A wall of schist is exposed along the north bank of Fisherman's Cove. This particular outcrop shows the schist has been tightly folded.

2 Outcropping of trondhjemite just off trail.

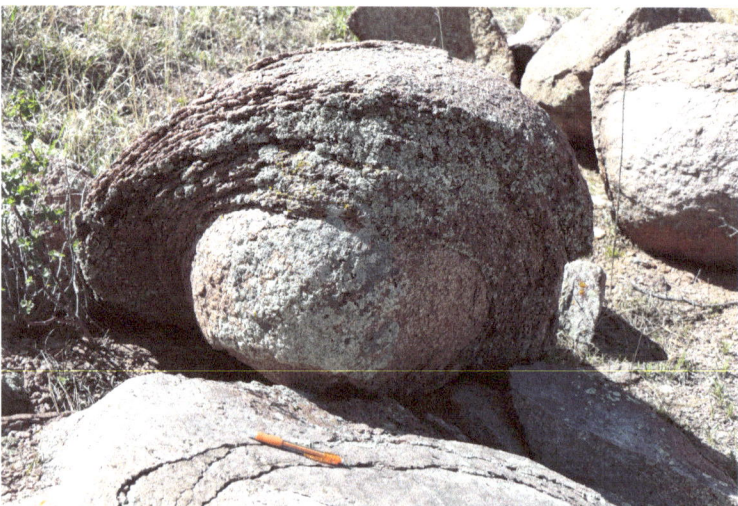

3 Great example of spheroidal weathering of the trondhjemite

4a The Great Unconformity is revealed by a 1.5 billion year break in the rock record. Here, the 300 million year old Fountain Fm on the right lays directly on top of 1.8 billion year old schist. At the far left, light colored trondhjemite is in contact with the schist.

4b Close-up view of the Fountain Fm overlying schist.

4c A fuller look at the Fountain Fm along the NW facing wall of the headland.

4d Boulder size fragments of eroded trondhjemite are part of the Fountain Fm deposit.

5 Exposure of schist with pods of blocky white quartz. The quartz is a product metamorphic reaction. Fragments of the quartz that have weathered out of the schist are common along the trail.

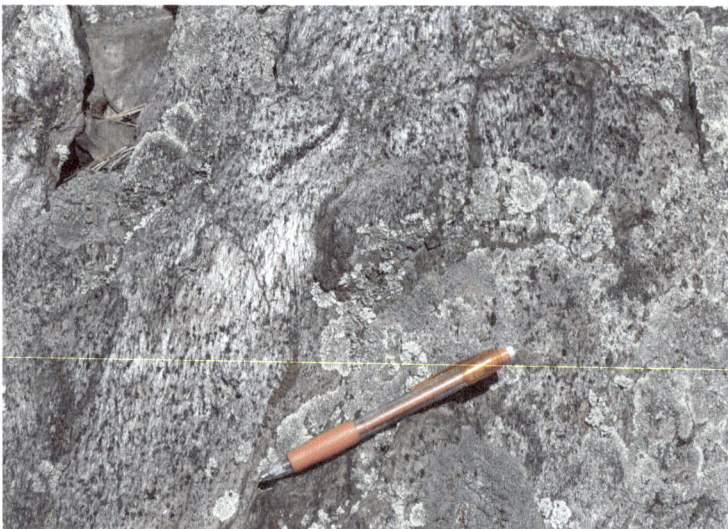

6 Schist outcrop with small black staurolite crystals. Staurolite is a metamorphic index mineral that indicates a relatively moderate grade of metamorphism.

7 Outcropping of meta-sandstone.

8 Another look at metamorphic reaction quartz. In this case the quartz is a purple-gray color.

9 Very large crystals of staurolite that appear to have random orientations with respect to the foliation. This type of schist with large mineral growths is commonly called a knotted mica schist.

View looking south over Pinewood Reservoir toward heavily wooded Blue Mountain—mostly composed of metamorphic gneiss.

HIKE #12
GATEWAY NATURAL AREA
Black Powder Trail
(2.2 miles & 450 feet of ascent—roundtrip)

The Black Powder Trail at Gateway offers a unique opportunity among Larimer County natural areas to view exposures of high grade metamorphic rocks. Mica schists bearing key index minerals, gneissic textures, and evidence of partial melting all indicate rocks created at depths of about 15 miles under high pressures and temperatures. This extreme environment developed within the core of an ancient mountain chain formed by the collision of landmasses at a convergent plate boundary ~ 1.8 billion years ago. Metamorphic rocks such as these were later intruded by igneous rocks and together form the foundational crust of Colorado. (For more geologic background, refer to section on Precambrian Crust in Part A of this guide)

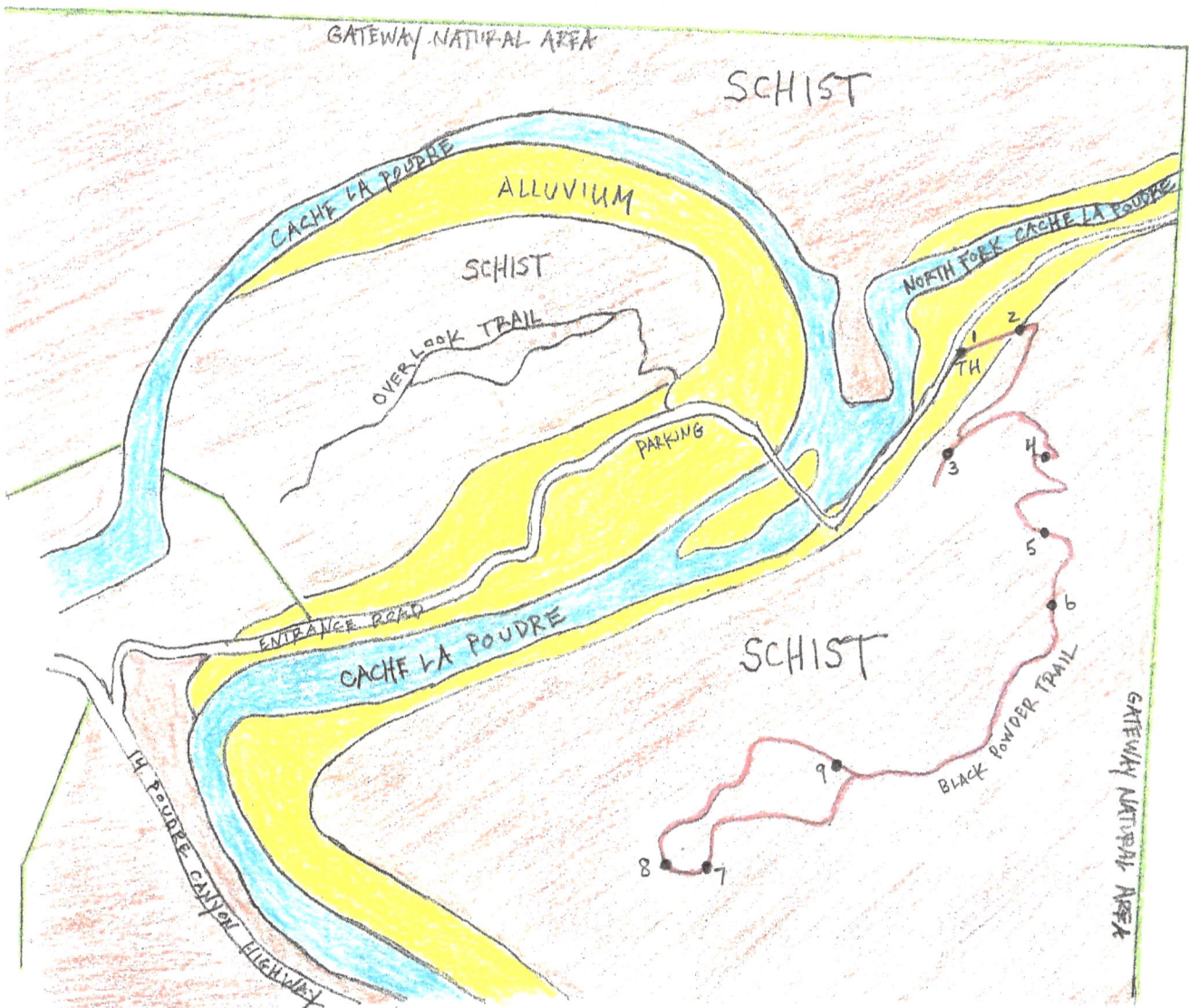

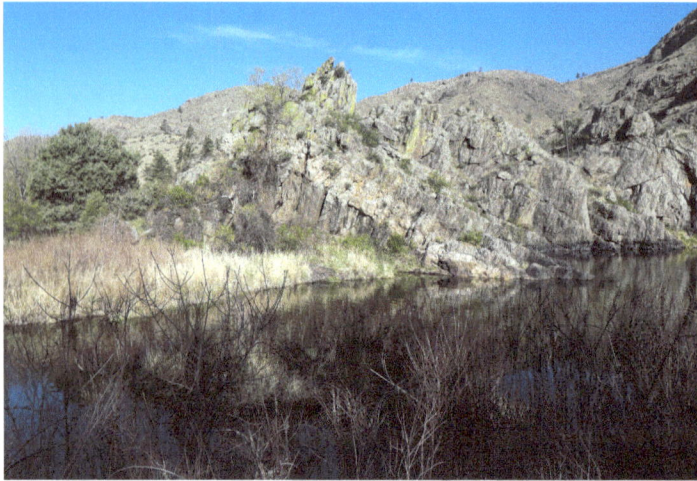

1 Looking across the north fork of the Cache La Poudre River from the trailhead, a light colored pegmatite dike is seen cutting through the host schist.

2 Schist comes in many varieties. On this hike the primary type is called a quartzofeldspathic schist. It has a quasi-banded texture and probably represents a transitional phase between schist and gneiss. With increased burial depth, mica in the schist becomes unstable and changes into the light colored feldspar lenses seen in this outcrop.

3a Along the short spur trail is an excellent exposure of a light colored pegmatite dike (upper half of outcrop) cutting through the schist.

3b Close-up shot of the pegmatite showing large milky white quartz, pink feldspars, and booklets of black biotite mica (above pencil).

4 A stepping stone along the trail showing the appearance of elongate white sillimanite crystals within the schist. Sillimanite is a metamorphic index mineral indicating deep burial conditions.

5 Another trailside block of schist exhibiting sillimanite as well as oxidized garnet crystals.

6 At even deeper burial conditions the schist starts to experience partial melting forming what is called a migmatite—a rock that is part igneous, part metamorphic.

7 Discontinuous lenses of blocky quartz are also commonly found within schist. Quartz is a also a product of metamorphic reaction.

8 Note the folds within this block of schist probably indicating the ductility of the rock at near melting temperatures.

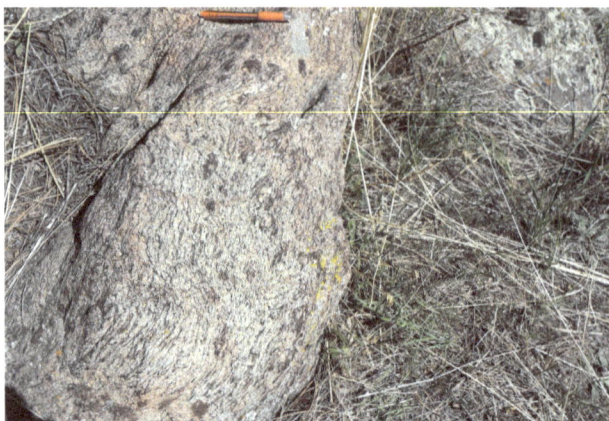

9 A boulder of quartzofeldspathic schist exhibiting knots of oxidized garnet minerals.

GLOSSARY

accreted terrane Slivers of **crust**, or microcontinents, that have been added onto the edge of an older and larger continent. *(Colorado is an accreted terrane that was added onto the southern margin of the older Wyoming craton)*

alabaster A fine-grained form of **gypsum** that is easy to carve.

alluvial fan A fan-shaped accumulation of sediment deposited by a stream as it exits a steep mountain canyon and flows onto flatter plains.

alluvium Sediments deposited by streams.

ammonites A group of marine mollusks noted for their spirally coiled shells. They became extinct with the dinosaurs at the end of the Cretaceous Period. *(commonly found in the Pierre Shale)*

Ancestral Rocky Mountains Ancient mountain ranges that formed 320 - 280 million years ago and were approximately located near today's Rocky Mountains.

antecedent stream a stream that predates uplifting of a mountain range and kept to its established course by downcutting as the mountain was uplifted.

anticline An arch shaped fold of rock layers. The oldest rocks are in the core of the fold.

asthenosphere Ductile zone of the upper mantle underlying the more rigid lithosphere.

basement Metamorphic and igneous rocks on which younger sedimentary rocks were deposited. Basement can be exposed when sedimentary rocks have been stripped off by erosion.

bedding plane Planar surface separating one layer (bed) of rock from another.

bedrock The solid rock underlying the cover of surface soil or loose sediment.

bentonite Clay formed by the weathering of volcanic ash.

braided stream A stream that simultaneously flows in a number of channels separated by bars or islands.

breccia A rock composed of angular rock fragments in a finer grained matrix.

butte An isolated topographic high with a flat top and steep sides.

cementation Part of the rock forming process that refers to the precipitation from groundwater of minerals (eg, quartz or calcite) that fill the pore spaces between sedimentary grains to form rock.

chalk A pure fine-grained skeletal limestone consisting of the remains of microscopic marine plants (algae) and animals (forams).

clastic sediment Composed of fragments eroded from pre-existing rock.

claystone A sedimentary rock composed chiefly of clay minerals.

coccolithophore Microscopic single-celled marine algae covered with calcareous plates called coccoliths.

concretion A hardened rock mass, commonly spherical, embedded within softer rock that formed as a result of focused precipitation of a dissolved mineral such as limestone, around some sort of central nucleus, commonly a fossil or fossil fragment.

conglomerate A sedimentary rock composed of rounded gravel set in a finer sand or silt sized matrix.

continental collision Collision between continental parts of two plates after the intervening oceanic part of the subducting plate has been completed.

continental crust The outermost shell of the earth forming the continents.

convection current Vertical circulation loop where warmer less dense rock rises and cooler denser rock sinks.

convergent plate boundary A boundary where two plates are moving toward each other. This type of boundary is either associated with subduction, where one plate slides beneath the other, or continental collision associated with accretion (growth of continents) and mountain building.

craton An old stable part of continental crust no longer affected by tectonic activity.

cross bedding Layers of sedimentary rock deposited at an angle to the main horizontal layering. Formed by migrating dunes created by wind or water currents. *(Commonly seen in the Fountain, Ingleside and Lyons rock units)*

cross section. A two dimensional portrayal of the structure of rock units below the earth's surface.

crust The outermost layer of earth. There are two types; continental and oceanic. continental crust is thicker and less dense, floating higher than oceanic crust which forms earth's ocean basins.

cyanobacteria Photosynthetic single-celled life form, still extant today, that have been found in **stromatolite** rocks over three billion years old. Thought to be the first oxygen-producing organisms that generated the oxygen in earth's atmosphere.

Denver Basin A topographic low adjacent to the Front Range that subsided as the Rocky Mountains grew. The sedimentary rocks that outcrop in the foothills form the western margin of the Denver Basin.

depositional environment The site and conditions under which sediment is deposited.

differential erosion Refers to variable rates of rock weathering due to changes in rock composition. *(In the foothills, the Dakota, Lyons and Ingleside are composed of resistant layers of quartz sandstone that result in topographically high ridges, or **hogbacks**)*

differential pressure An unequal force exerted on or by a rock in a particular direction. For example, when two continents collide a horizontal force is dominant.

dike A tabular igneous intrusion that cuts across bedding or foliation.

dip The angle at which a rock layer is tilted, measured from the horizontal.

dome An anticlinal structure that plunges equally in all directions.

dowsing (aka water-witching or divining) The use of hand-held copper wires or sticks to detect the presence of groundwater. Not a scientifically proven method.

eolian Term referring to the sedimentary processes or deposits formed by wind.

erosion The removal and transport of rock and soil by moving wind, water and ice.

escarpment A cliff or steep slope bounding a topographic high.

feldspar The most abundant rock-forming mineral group. The group is split into the *alkali feldspars* with varying proportions of potassium and sodium in their chemical formula, and the *plagioclase feldspars* with varying proportions of sodium and calcium in their chemical formula.

flatiron Triangular landform created by erosional breaches through a steeply dipping hogback or along the steeply dipping beds of an anticline.

fold A curvilinear bend in layers of rock.

fold-thrust belt A mountain belt of thrust faults and related folds developed in response to continental collision and crustal shortening.

foliation A fabric developed in some metamorphic rocks that reflects the parallel alignment of platy minerals (e.g. mica schist) or compositional banding (gneiss).

foreland basin A basin formed adjacent to a thrust belt mountain front where stacked thrust sheets cause the crust to bend downwards and subside. *(Colorado was part of an extensive foreland basin formed in front of the thrust belt mountains in central Utah during the Sevier Orogeny. Subsequent uplifts associated with growth of the Rocky Mountains compartmentalized the foreland basin into individual basins such as the Denver Basin bordering the Front Range)*

formation A recognizable interval of rock that can be mapped over a fairly broad region and is confined to a definable interval of time.

Front Range Part of the southern Rocky Mountains, the Front Range is the first range of mountains rising up from the Great Plains, extending from southern Wyoming to Pueblo, Colorado.

gneiss A high-grade **metamorphic rock** with distinctive alternating bands of light and dark colored minerals.

granite A light colored intrusive igneous rock dominantly composed of quartz and feldspar with added mica and hornblende as secondary components.

granitic General term referring to all light colored, granite like igneous rocks.

granodiorite An intrusive igneous rock similar to granite but with less quartz and more sodium feldspar content. Typically light to dark gray in color with a salt and pepper appearance.

Great Unconformity The surface that represents a gap in the geologic rock record separating igneous and metamorphic basement rocks circa 1.8 billion years old from overlying sedimentary rocks circa 300 million years old (in the northern Colorado foothills).

group A rock unit made up of more than one formation.

gypsum A soft evaporite mineral ($CaSO_4.2H_2O$) occurring as deposits formed by evaporating sea water.

headward erosion The process by which a stream lengthens by eroding upslope.

head scarp The crescent shaped detachment surface at the top of a rockslide.

hogback Refers to the steep sided ridge formed by a steeply dipping erosion resistant sedimentary rock. *(In the foothills the hogbacks are formed by the Dakota, Lyons and Ingleside rock units)*

hydrothermal fluids Hot subsurface water usually enriched in dissolved minerals and associated with mineral deposits.

igneous rock A rock formed by solidification of magma.

index minerals Minerals that serve as indicators of the degree of metamorphism.

inoceramous An extinct genus of clams. *(Common to the Niobrara Formation)*

intrusion A rock formed by cooling and hardening of magma that was emplaced within older rock beneath the earth's surface.

inverted topography When valleys or synclines, normally topographic low features, evolve through the effect of differential erosion to become topographic high features in a landscape.

lamination Fine scale layering within a rock, less than 1 cm in thickness.

Laramide Orogeny The mountain building event that formed the current Rocky Mountains between 70 and 40 million years ago.

leach To dissolve away.

limestone A sedimentary rock composed mostly of the mineral calcite ($CaCO_3$) and mainly formed by accumulation of the skeletal remains of marine organisms.

lithology Rock type; eg sandstone, shale, granite, schist etc.

lithosphere Rigid outer layer of rocks that comprise the mobile tectonic plates overlying the ductile asthenosphere. Includes the crust plus upper part of the mantle.

mantle The layer of earth between the crust and core. Occupies 84% of earth's volume.

marlstone A silty, clayey limestone.

member A particular recognized layer of rock within a formation.

metamorphic rock A rock whose original mineralogy and texture has been changed by the effects of heat, pressure and/or fluids.

mica A group of minerals that have a thin sheet-like structure. The dominant types are black colored *biotite,* and clear to light brown colored *muscovite.* Both are highly reflective in sunlight.

microcrystalline Crystals too small to be seen with the naked eye. In a limestone, calcite crystals smaller than 4 micrometers in size.

migmatite A high grade metamorphic rock that has been altered by the inclusion of igneous minerals (e.g. quartz and feldspar) formed by partial melting of the rock.

mudstone Similar to shale, composed of clay and silt sized material but does not split into thin layers.

oceanic crust The outermost shell of the earth forming the ocean floors.

orogeny A mountain building event, especially those formed by colliding continents.

outcrop An exposed rock unit at the earth's surface.

paleosol A preserved ancient soil layer within the rock record.

Pangaea An ancient supercontinent that developed starting around 300 million years ago and began breaking apart 200 million years ago to form today's continents.

pegmatite An intrusive igneous rock, usually of granitic composition, having very large mineral crystals (> 1 inch).

permeability The ability of a rock to allow fluids to flow through it.

plate tectonics The geologic theory that the outer layer of the earth is composed of rigid plates of rock that move with respect to each other. Their interaction at plate boundaries is the cause of earthquakes, mountain building, volcanism and the collision and growth of continents.

pluton General term for an intrusive igneous rock regardless of size, shape or composition.

porosity The percentage of void space in a rock that is filled by water, oil or gas.

Precambrian Interval of geologic time from earth's formation roughly 4.5 billion years ago up until the Cambrian Period about 540 million years ago which began a great diversification of life forms.

quartz A common rock-forming mineral composed of silicon dioxide (SiO_2). Typically white or clear in color, but the presence of trace elements can reflect other colors as well.

red beds Sedimentary rocks that have been colored red by a coating of the mineral hematite (Fe_2O_3).

rockslide Downslope movement of cohesive slabs of rock.

rock unit A general term referring to a discrete body of rock without regard to its classification status as a group, formation or member.

sandstone A sedimentary rock composed of sand-sized grains (1/16 to 2 mm).

schist A metamorphic rock composed dominantly of parallel aligned platy minerals. Commonly formed by the metamorphosis of clay minerals in a shale into mica.

sea-level cycles The rise and fall of sea levels often associated with fluctuating periods of glaciation, or changes in ocean basin volume caused by subsea volcanism. A primary control on marine depositional environments.

sedimentary rock Rock formed by sediments accumulated at the earth's surface as either material eroded from pre-existing rock (eg sandstone), or from biological remains (eg limestone), or as a chemical precipitate (eg gypsum).

Sevier Orogeny A mountain building event affecting western North America between 150—65 million years ago.

shale A fine-grained sedimentary rock composed of clay and silt that typically splits along thin bedding parallel layers.

shear zone A narrow region associated with continental collision that has been severely ductilely deformed.

sill A tabular igneous body intruded between layers of sedimentary rock or parallel to foliation in metamorphic rock.

siltstone A sedimentary rock composed of silt-sized grains (1/156 to 1/16 mm).

slickensides Lineations on a fault or bedding plane caused by frictional sliding of one rock body against another.

source rock A shale or limestone containing organic matter capable of producing oil and gas.

spheroidal weathering Commonly developed in granitic rocks exposed at the surface where enhanced weathering occurs at joint intersections. With overall surface weathering and expansion the rock develops concentric shells resembling an onion-skin appearance.

stratigraphic chart A diagram illustrating the layered sequence of rocks from oldest to youngest in a particular area of region.

stream capture Also known as stream piracy, the process where one stream captures another's flow through headward erosion.

strike valley A valley that parallels the mapped surface orientation of a rock unit.

stromatolite A laminated, mounded rock that was built up by successive layers of cyanobacteria that trapped sediments as they grew up in the water column.

subduction zone the region where a dense oceanic plate descends into the asthenosphere at a convergent plate boundary.

supercontinent A giant, ancient continent that through many collisions and accretions was an assembled collection of all or most of the continents at a given time.

superposed stream A stream that has cut down into an older structural landscape but maintains its established course; therefore possibly cutting through old buried anticlines.

suture zone The zone demarcating where two continents collided to form a singular landmass.

syncline A trough shaped fold of rock layers. The youngest rocks are in the core of the fold.

thrust fault An inclined surface where a block of overlying rock moves up and over a block of underlying rock.

toe thrust A thrust fault developed at the foot of a rockslide accommodating detachment at the base of a slope failure.

tonalite A type of granitic rock with the feldspar content being >90% *plagioclase.*

trace fossil An imprint left by an organism in sediment that became preserved in rock. Footprints and burrows are examples.

trondhjemite A light colored variety of tonalite with feldspars having more sodium content, and darker colored minerals like biotite and hornblende making up less than 10% of the rock.

unconformity A gap in the rock record represented by a surface of contact between two rock units indicating a period of erosion and cessation of sedimentation.

volcanic arc A magmatic zone developed above a subducted plate. If the upper plate is composed of oceanic crust, then the string of developed volcanoes is called an island arc.

volcanic ash Formed by a gas-charged spray of magma exploded from a volcano that quickly chills into tiny glass particles.

weathering The disintegration and decomposition of rock by physical and chemical processes at the earth's surface.

Western Interior Seaway An inland sea that stretched from the Gulf of Mexico to the Arctic, flooding Colorado from about 100 to 70 million years ago.

Yavapai Orogeny A mountain building event that occurred about 1.8 billion years ago marking the collision of landmasses that accreted Colorado **crust** to the Wyoming Craton.

GENERAL REFERENCES

GEOLOGY OF COLORADO - POPULAR

Abbott L, Cook T (2012) Geology Underfoot Along Colorado's Front Range. Mountain Press, Missoula, Montana.

Baldridge SW (2004) Geology of the American Southwest. Cambridge.

Blakey RC, Ranney WD (2018) Ancient Landscapes of Western North America. Springer Nature.

Chronic J, Chronic H (1972) Prairie Peak and Plateau, A Guide to the Geology of Colorado. Colorado Geological Survey Bulletin 32.

Chronic H, Williams F (2014) Roadside Geology of Colorado, Third Edition. Mountain Press, Missoula, Montana.

Colorado Stratigraphy. https://www.coloradostratigraphy.org

Matthews V et al. (eds) (2003) Messages in Stone - Colorado's Colorful Geology. Colorado Geological Survey.

U.S. GEOLOGICAL SURVEY, GEOLOGIC QUADRANGLE MAPS

Braddock WA et al. (1989) Geologic Map of the Horsetooth Reservoir Quadrangle, Larimer County, Colorado. U.S. Geological Survey Map GQ-1625.

Braddock WA et al. (1988) Geologic Map of the Laporte Quadrangle, Larimer County, Colorado. U.S. Geological Survey Map GQ-1621.

Braddock WA et al. (1988) Geologic Map of the Livermore Quadrangle, Larimer County, Colorado. U.S. Geological Survey Map GQ-1618.

Braddock WA et al. (1970) Geologic Map of the Masonville Quadrangle, Larimer County, Colorado. U.S. Geological Survey Map GQ-832.

Braddock WA et al. (1988) Geologic Map of the Table Mountain Quadrangle, Larimer County, Colorado. U.S. Geological Survey Map I-1805.

Punongbayan R et al. (1989) Geologic Map of the Pinewood Lake Quadrangle, Boulder and Larimer Counties, Colorado. U.S. Geological Survey Map GQ-1627.

Workman JB et al. (2018) Geologic Map of the Fort Collins 30'x60' Quadrangle, Larimer and Jackson Counties, Colorado, and Albany and Laramie Counties, Wyoming. U.S. Geological Survey Scientific Investigations Map 3399.

TECHNICAL REFERENCES

PRECAMBRIAN CRUST

Houston RS et al. (1989) A Review of the Geology and Structure of the Cheyenne Belt and Proterozoic Rocks of Southern Wyoming. Geologic Society of America Paper 235.

Mahon KH et al. (2013) Proterozoic Metamorphism and Deformation in the Northern Colorado Front Range. The Geological Society of America, Field Guide 33.

Moyen JF, Martin H (2012) Forty Years of TTG Research. Lithos 148: 312-336.

Polat A (2012) Growth of Archean Continental Crust in Oceanic Island Arcs. Geology 40/4 383-384.

Selverstone J et al. (1997) Proterozoic Tectonics of the Northern Colorado Front Range. Rocky Mountain Association of Geologists, Colorado Front Range Guidebook.

FOUNTAIN FORMATION

Chapin CE et al. (2014) The Rocky Mountain Front, Southwestern USA. Geosphere 10/5: 1043-1060.

Crick W, Ethridge FG (1985) Pennsylvanian and Permian Sedimentary Rocks, Horsetooth Reservoir. Colorado State University.

Gorvett Z (2021) The Strange Race to Track Down a Missing Billion Years. BBC Future, The Lost Index | Geology (published September 1, 2021).

Hogan I (2013) Paleo-Fluid Migration and Diagenesis in the Pennsylvanian-Permian Fountain Formation. Colorado State University, MS Thesis.

Howard JD (1966) Patterns of Sediment Dispersal in the Fountain Formation of Colorado. The Mountain Geologist 3/4: 147-153.

Huntoon JE et al. (2014) Fossil Plants from a Proximal Alluvial-Fan Complex: Implications for Late Paleozoic Sedimentary Processes in Western Tropical Pangea. Utah Geological Association Publication 43: 473-490.

Leary RJ et al. (2017) A Three-Sided Orogen: A New Tectonic Model for Ancestral Rocky Mountain Uplift and Basin Development. Geology 45/8: 735-738.

Marshak S et al. (2000) Inversion of Proterozoic Extensional Faults: An Explanation for the Pattern of Laramide and Ancestral Rockies Intracratonic Deformation. Geology 28/8: 735-738.

Meyer AJ (2007) Variations Among Paleosol Characteristics in Relation to Fluvial Deposits, Pennsylvanian Fountain Formation, Northern Colorado Front Range. Colorado State University, MS Thesis.

Napp KF, Ethridge FG (1985) Depositional Systems of Fountain Formation and
    its Basinal Equivalents, Northwestern Denver Basin, Colorado. American Association of
    Petroleum Geologists Bulletin 69, Meeting Abstract.

Sturmer DM et al. (2018) Tectonic Analysis of the Pennsylvanian Ely-Bird Spring Basin: Late
    Paleozoic Tectonism on the Southwestern Laurentia Margin and the Distal Limit of the
    Ancestral Rocky Mountains. Tectonics 37: 604-620.

Sweet DE et al. (2015) Proposing an Entirely Pennsylvanian Age for the Fountain
    Formation through New Lithostratigraphic Correlation along the Front Range.
    The Mountain Geologist 52/2: 43-70.

INGLESIDE FORMATION

Maughan EK et al. (1985) Pennsylvanian and Permian Eolian Sandstone Facies, Northern
    Colorado and Southeastern Wyoming. The Rocky Mountain Section SEPM Field Trip
    Guide: Many Ways to Future Plays.

Nair K (2018) Facies Reconstruction and Detrital Zircon Geochronology of the Ingleside/
    Casper Formation. Colorado State University, MS Thesis.

OWL CANYON FORMATION

Al-Ani RAB (1992) Sedimentology of the Owl Canyon Formation between the Colorado-
    Wyoming State Line and Lyons, Colorado. Colorado State University, PhD.
    Dissertation.

Larson SM (2009) Laramide Transpression and Oblique Thrusting in the Northeastern
    Front Range, Colorado: 3D Kinematics of the Livermore Embayment. Colorado
    State University, MS Thesis.

Maughan EK, Wilson RF (1960) Pennsylvanian and Permian Strata in Southern Wyoming
    and Northern Colorado. U.S. Geological Survey, Guide to the Geology of Colorado.

LYONS SANDSTONE

Adams J, Patton J (1979) Sebkha-Dune Deposition in the Lyons Formation (Permian)
    Northern Front Range, Colorado. The Mountain Geologist 16/2: 47-57.

Stone DS (1985) Seismic Profiles in the area of the Pierce and Black Hollow Fields, Weld
    County, Colorado. Rocky Mountain Association of Geologists.

Walker TR, Harms JC (1972) Eolian Origin of Flagstone Beds, Lyons Sandstone (Permian)
    Type Area, Boulder County, Colorado. The Mountain Geologist 9/2-3: 279-288.

LYKINS FORMATION

Hagadorn JW et al. (2016) The Permian-Triassic Transition in Colorado. Geological Society of America Field Guide 44: 73-92.

Hagadorn JW et al. (2019) Identifying a Mass Extinction in Front Range Open Space: Age and Environments of the Lykins Formation. Department of Earth Sciences, Denver Museum of Nature and Science.

Johnson KS (2008) Gypsum-Karst Problems in Constructing Dams in the USA. Environmental Geology 53: 945-950.

Story JA, Howell DH (1963) Gypsum Deposits of the Northern Colorado Front Range. Rocky Mountain Association of Geologists, Geology of the Northern Denver Basin and Adjacent Uplifts.

Warren NR (2016) Comparative Analysis of Stromatolite Dome Spacing and Grazing Levels in the Lykins Formation of Colorado and Southern Wyoming. University of Colorado, Boulder, MS Thesis.

JELM and SUNDANCE FORMATIONS

Blakey RC et al. (1988) Synthesis of late Paleozoic and Mesozoic Eolian Deposits of the Western Interior of the United States. Sedimentary Geology 56: 3-125.

Reeside JB (1931) Supposed Marine Jurassic (Sundance) in Foothills of Front Range Colorado. U.S. Geological Survey, Geological Notes.

Reeside JB (1929) Triassic-Jurassic 'Red Beds' of the Rocky Mountain Region: A Discussion. Journal of Geology 37/1: 47-63.

MORRISON FORMATION

Burnell J (2011) Colorado's Uranium Deposits. Colorado Geological Survey.

Chenoweth WL (1981) The Uranium-Vanadium Deposits of the Uravan Mineral Belt and Adjacent Areas, Colorado and Utah. New Mexico Geological Society Guidebook, 32nd Field Conference, Western Slope Colorado.

Dye L (1990) Bones of Rare Dinosaur Discovered in Colorado. Los Angeles Times January 4, 1990.

Maidment SCR, Muxworthy A (2019) A Chronostratigraphic Framework for the Upper Jurassic Morrison Formation, Western USA. Journal of Sedimentary Research 89/10: 1017-1038.

Turner CE et al. (2004) Reconstruction of the Upper Jurassic Morrison Formation Extinct Ecosystem - a Synthesis. Sedimentary Geology 167: 309-355.

Wilkin J (2019) Patterns in Paleontology: the Real Jurassic Park. palaeontologyonline.com.

DAKOTA GROUP

Braddock WA (1978) Dakota Group Rockslides, Northern Front Range, Colorado,USA. Elsevier, Developments in Geotechnical Engineering 14: 439-475.

Braddock WA, Eicher DL (1962) Block-Glide Landslides in the Dakota Group of the Front Range Foothills, Colorado. Geological Society of America Bulletin 73: 317-324.

Dolson J et al. (1991) Regional Paleotopographic Trends and Production, Muddy Sandstone (Lower Cretaceous), Central and Northern Rocky Mountains. American Association of Petroleum Geologists Bulletin 75/3: 409-435.

Higley D et al. (2003) Petroleum System and Production Characteristics of the Muddy (J) Sandstone (Lower Cretaceous) Wattenberg Continuous Gas Field, Denver Basin, Colorado. American Association of Petroleum Geologists Bulletin 87/1:15-37.

MacKenzie DB (1965) Depositional Environments of Muddy Sandstone, Western Denver Basin, Colorado. American Association of Petroleum Geologists Bulletin 49/2: 186-206.

Waage KM (1955) Dakota Group in Northern Front Range Foothills, Colorado. Geological Survey Professional Paper 274-3.

BENTON GROUP

Durkee HM (2016) Reservoir Characterization and Geochemical Evaluation of the Greenhorn Formation in the Northern Denver Basin, Colorado. Colorado School of Mines, MS Thesis.

Larson RL (1995) The Mid-Cretaceous Superplume Episode. Scientific American 272/2: 82-86.

Pang M (1995) Tectonic Subsidence of the Cretaceous Western Interior Basin, United States. Louisiana State University, PhD Dissertation.

Turgeon SC, Creaser RA (2008) Cretaceous Oceanic Anoxic Event 2 Triggered by a Massive Magmatic Episode. Nature 454: 323-325.

NIOBRARA FORMATION

Public Lands History Center, Concrete and Cement. publiclands.colostate.edu

O'Neal DL (2015) Chemostratigraphic and Depositional Characterization of the Niobrara Formation, Cemex Quarry, Lyons, Colorado. Colorado School of Mines, MS Thesis.

Wood JB (2017) Regional Geology and Chemostratigraphy of the Fort Hays Member of the Niobrara Formation, Western Interior, USA. Colorado School of Mines, MS Thesis.

PIERRE SHALE

Ball MW (1924) Geological Notes - Gas Near Fort Collins, Colorado. American Association of Petroleum Geologists Bulletin 8: 79-87.

Bertog J (2010) Stratigraphy of the Lower Pierre Shale (Campanian): Implications for the Tectonic and Eustatic Controls on Facies Distributions. Journal of Geological Research.

Scott GR, Cobban WA (1986) Geologic, Biostratigraphic, and Structure Map of the Pierre Shale between Loveland and Round Butte, Colorado. USGS Miscellaneous Investigations Series Map I-1700.

Scott GR, Cobban WA (1959) So-Called Hygiene Group of Northeastern Colorado. Rocky Mountain Association of Geologists, Symposium on Cretaceous Rocks of Colorado.

LARAMIDE OROGENY

Axen GJ (2018) Basal Continental Mantle Lithosphere Displaced by Flat-Slab Subduction. Nature Geoscience 11: 961-964.

Dickinson WR et al. (1988) Paleogeographic and Paleotectonic Setting of Laramide Sedimentary Basins in the Central Rocky Mountain Region. Geological Society of America Bulletin 100: 1023-39.

English JM, Johnston ST (2004) The Laramide Orogeny: What Were the Driving Forces? International Geology Review 46: 833-38.

Gutscher MA (2018) Scraped by Flat-Slab Subduction. Nature Geoscience 11: 890-91.

Sun D et al. (2017) A Dipping Thick Segment of the Farallon Slab Beneath Central U.S. Journal of Geophysical Research: Solid Earth 122: 2911-28.

SOAPSTONE PRAIRIE

Bryan K, Louis RL (1940) Geologic Antiquity of the Lindenmeier Site in Colorado. Smithsonian Miscellaneous Collections 99/2.

Labell JM (2016) Lindenmeier Folsom Site. Colorado Encyclopedia. Accessed February 16, 2022. https://coloradoencyclopedia.org/article/lindenmeier-folsom-site

Leonard EM (2002) Geomorphic and Tectonic Forcing of Late Cenozoic Warping of the Colorado Piedmont. Geologic Society of America 30/7: 595-598.

Little JB (2009) The Ogallala Acquifer: Saving a Vital U.S. Water Source. Scientific American Special Edition 19: 32-39.

Moore FE (1959) The Geomorphic Evolution of the East Flank of the Laramie Range, Colorado and Wyoming. University of Wyoming, PhD Thesis.

Price WA et al. (1946) Algae Reefs in Cap Rock of Ogallala Formation on Llano Estacado Plateau, New Mexico and Texas. American Association of Petroleum Geologists 30/10: 1742-1746.

Willlett S et al. (2018) Transience of the North American High Plains Landscape and its Impact On Surface Water. Nature 561: 528-532.

FIGURE CREDITS

INTRODUCTION

PRECAMBRIAN CRUST

FOUNTAIN FORMATION

INGLESIDE FORMATION

OWL CANYON FORMATION

BENTON GROUP

NIOBRARA FORMATION

PIERRE SHALE

LARAMIDE OROGENY

SOAPSTONE PRAIRE

NATURAL AREA HIKES

(All figures and photos by author)

About the Author

Mike's fascination with geology began as a young child in his grandfather's garden in Chicago, which was outlined with a variety of interesting rocks he had collected on his travels to each of the lower 48 states. And summer fishing trips with him to the north woods of Minnesota, Michigan, and Wisconsin usually included a trip to a nearby nature center, where the geology hook was set further. Exhibits of rocks sometimes billions of years old left a lasting impression. It was there that he first learned that the earth was decipherable and its history could be interpreted and understood. And beneath the thin layer of surface soil many of us only get to know lays a mysterious world of rocks, the fundamental substance upon which all else is built, with a story to tell.

Mike earned his BS in Geology from University of Illinois in Chicago and his MS in Geology from Wright State University. Mike spent 33 years interpreting subsurface geology pursuing oil and gas resources in the US and internationally. Upon retirement in 2018, Mike and his wife relocated to Fort Collins, Colorado—a city adjacent to foothills and mountains with a wealth of open space natural areas to explore.

Mike is currently a volunteer naturalist with the City of Fort Collins Natural Areas Department, and a volunteer hike leader with the Fort Collins Newcomers Club. He is a member of The Geological Society of America, Rocky Mountain Association of Geologists, and Northern Colorado Geologists.